Newton's Third Rule and the Experimental Argument for Universal Gravity

This book provides a reading of Newton's argument for universal gravity that is focused on the evidence-based, "experimental" reasoning that Newton associates with his program of experimental philosophy. It highlights the richness and complexity of the *Principia* and also draws important lessons about how to situate Newton in his natural philosophical context.

The book has two primary objectives. First, it defends a novel interpretation of the third of Newton's four Rules for the Study of Natural Philosophy – what the author terms the Two-Set Reading of Rule 3. Second, it argues that this novel interpretation of Rule 3 sheds additional light on the differences between Newton's experimental philosophy and Descartes's "hypothetical philosophy," and that it also illuminates how the practice of experimental philosophy allowed Newton to make a universal force of gravity the centerpiece of his explanation of the system of the world.

Newton's Third Rule and the Experimental Argument for Universal Gravity will be of interest to researchers and advanced students working on Newton's natural philosophy, early modern philosophy, and the history of science.

Mary Domski is Professor of Philosophy at the University of New Mexico. Her research focuses on the philosophy, mathematics, and science of the early modern period. She has authored numerous papers on Newton and Descartes and edited the special issue "Newton and Newtonianism" for *The Southern Journal of Philosophy* (2012).

Routledge Focus on Philosophy

Routledge Focus on Philosophy is an exciting and innovative new series, capturing and disseminating some of the best and most exciting new research in philosophy in short book form. Peer reviewed and at a maximum of fifty thousand words shorter than the typical research monograph, *Routledge Focus on Philosophy* titles are available in both ebook and print-on-demand format. Tackling big topics in a digestible format the series opens up important philosophical research for a wider audience, and as such is invaluable reading for the scholar, researcher and student seeking to keep their finger on the pulse of the discipline. The series also reflects the growing interdisciplinarity within philosophy and will be of interest to those in related disciplines across the humanities and social sciences.

The Right to Know
Epistemic Rights and Why We Need Them
Lani Watson

Honouring and Admiring the Immoral
An Ethical Guide
Alfred Archer and Benjamin Matheson

Newton's Third Rule and the Experimental Argument for Universal Gravity
Mary Domski

For more information about this series, please visit: www.routledge.com/Routledge-Focus-on-Philosophy/book-series/RFP

Newton's Third Rule and the Experimental Argument for Universal Gravity

Mary Domski

Routledge
Taylor & Francis Group

NEW YORK AND LONDON

First published 2022
by Routledge
605 Third Avenue, New York, NY 10158

and by Routledge
2 Park Square, Milton Park, Abingdon, Oxon, OX14 4RN

Routledge is an imprint of the Taylor & Francis Group, an informa business

© 2022 Mary Domski

Library of Congress Cataloging-in-Publication Data
A catalog record for this title has been requested

ISBN: 978-1-032-02036-5 (hbk)
ISBN: 978-1-032-02622-0 (pbk)
ISBN: 978-1-003-18425-6 (ebk)

DOI: 10.4324/9781003184256

Typeset in Times New Roman
by codeMantra

To Michael and Nico,
with infinite and absolute thanks

Contents

Preface and Acknowledgments

Readers less familiar with Newton's writings might find it surprising that a short book could be written about a single sentence of the *Principia*. Readers more familiar with Newton will likely be surprised that a book on Rule 3 is short. No question, the book could have been shorter, and it could have been longer. It stands as it does, because I had two primary aims in mind while I was writing. I wanted to provide a general narrative about the meaning and significance of Rule 3 that did justice to the complexity and richness of the *Principia*, and I wanted to strike a balance between offering an interpretation of the Rule that was persuasive and offering it in a way so that a wider audience could come along for the ride. I probably could have said less, and had I ventured farther beyond the published version of the *Principia*, I definitely could have said more. But I hope I have said enough to bring light to a worthwhile way of thinking about Rule 3 and about Newton's experimental philosophy.

This book would likely not exist if it weren't for the kind invitation that I received in 2018 to speak at the conference "Responses to Newton: The impact of the mathematical-experimental paradigm on natural philosophy, epistemology and metaphysics (1687–1800)." (The conference was held from 5 to 7 June 2019 at the Institute of Philosophy of KU Leuven and was funded through grants from The Research Foundation – Flanders (FWO) and The Fund for Scientific Research (FNRS).) Only after getting that invitation did I start to grapple with the meaning and significance of Rule 3, and I am incredibly grateful to Karin de Boer, Steffen Ducheyne, Stephen Howard, Arnaud Pelletier, and Anne-Lise Rey for organizing a terrific conference and giving me the opportunity to present my ideas on why Newton's contemporaries may have found the program of natural philosophy in the *Principia* so difficult to understand. All the feedback that I received from my audience members in Leuven was generous and constructive, and I am especially grateful to Karin, Steffen, Arnauld, Lisa Downing, Philippe Hamou, Dimitris Petakos, and Troy Vine for offering comments that helped me see what additional work I had to do.

Prior to the "Responses to Newton" conference, I presented a rough sketch of my interpretation of Rule 3 at the 2019 Du Châtelet Prize Workshop, which was organized by Katherine Brading and held at Duke University from 4 to 5 April 2019. The workshop gave me the opportunity to think out loud about Newton's presentation of Rule 3 in the third edition *Principia*, and I am grateful to Katherine, Andrew Janiak, Adwait Parker, Chris Smeenk, George Smith, and Marius Stan for their comments on how Newton's various remarks could be interpreted.

I completed the preliminary write-up of my Two-Set Reading in early 2020 and had pitched that draft as a scholarly article that covered a narrow selection of interpretive issues. Zvi Biener and Niccolò Guicciardini graciously agreed to read what I had put together, and it was because of their feedback that I began to see how I could couch my reading of Rule 3 in a broader narrative about the experimental philosophy that Newton pursues in Book 3. Zvi's comments were especially important in this regard. He pressed me on issues that I had not treated with adequate care, and in trying to address his questions I began to see connections between different parts of Book 3, and also between Newton's *Principia* and Descartes's *Principles*, that weren't initially on my radar.

The two anonymous reviewers of the manuscript also went above and beyond. Even if I have not adequately addressed all their concerns, the final manuscript is all the better because of the pages of thoughtful commentary that they supplied. For securing such exemplary reviewers, and for overseeing the entire publication process, I have Andrew Weckenmann to thank. Andrew, Allie Simmons, and the entire team at Routledge were terrific to work with, and I am grateful for the time that they invested in this project. I also owe thanks to Mark Peceny, the former and decade-long Dean of the University of New Mexico's College of Arts and Sciences. With Mark's permission, I used research funds that I received for serving as a Special Assistant in his office to make the book available in Open Access format.

Among the many people who helped me see this project to completion, none have been more important than my family and friends. My deepest, most heartfelt thanks go to my mom, Thuan; to my sister, Ann, and my nephew, Jackson; to Mary Kathryn Karafonda and Kathleen Karafonda; and to my dad, Dieter, who I still hear clapping for me from beyond the grave. Summoning the energy to think and write, especially during a year like 2020, would have been impossible without them.

Finding my way to the reading of the *Principia* that I present in this book would have been impossible were it not for a number of other people. For more than a decade, Zvi, Niccolò, Katherine, Andrew, Chris, Marius, Sarah Pessin, Don Rutherford, Eric Schliesser, and Ericka Tucker have given me feedback that has helped me better understand Newton's writings and better appreciate the philosophical context in which Newton was working. They also offered support and encouragement along the way, and I could not be more grateful.

As far as reading Newton goes, my greatest debt and deepest gratitude are owed to Michael Friedman and Nico Bertoloni Meli. It has now been over 20 years since I took their seminars on Newton at Indiana University. But it is because of what Michael and Nico taught me then that I continue to read the *Principia* as I do – historically, philosophically, and with due attention to details that would be easier left ignored. Their imprint can be found throughout the pages that follow, and dedicating this book to Michael and Nico seemed only fitting. It was also the very least that I could do.

1 Introduction

1.1 Gravity, Descartes, and Experimental Philosophy

Isaac Newton's *Principia* was first published in 1687, with second and third editions appearing in 1713 and 1726, respectively. Between each edition Newton made noticeable changes to the text.[1] However, several major elements remained unchanged. All three editions have the same full title: *Philosophiae Naturalis Principia Mathematica*, or *Mathematical Principles of Natural Philosophy*. All three editions also share the same general structure. The work opens with two sections (entitled "Definitions" and "Axioms, or the Laws of Motions") that are followed by three books. In all editions, Book 1 is dedicated to the study of motion in nonresisting media, Book 2 to the study of motion in resisting media, and Book 3 to an explanation of celestial motions in terms of a universal force of gravity. And in all editions, Book 2 ends with Newton branding his explanation of celestial motions as an alternative to and improvement over the explanation of these motions by the use of vortices:

> the hypothesis of vortices can in no way be reconciled with astro-nomical phenomena and serves less to clarify the celestial motions than to obscure them. But how those motions are performed in

1 The full range of changes between the three editions of the *Principia* is tracked in Newton (1999), the now-standard translation of the third edition produced by I. Bernard Cohen and Anne Whitman. In the *Guide* that precedes the transla-tion, Cohen (1999) details the historical context surrounding the more extensive revisions that Newton made between the various editions and provides citations to relevant secondary literature. Unless otherwise noted, all citations to English translations of the *Principia* in what follows are to Newton (1999), with bracketed terms and phrases adopted from that translation.

DOI: 10.4324/9781003184256-1

free spaces without vortices can be understood from book 1 and will be shown more fully in book 3 on the system of the world.

(Newton 1999, 790)

When the *Principia* was first published, the most well-known and influential version of the Vortex Hypothesis was the one presented by René Descartes in his *Principles of Philosophy* (1644).[2] The vortex explanation of planetary motion is laid out in Part III (entitled "Of the Visible Universe"), and it rests on two basic assumptions: [1] There is a subtle ethereal fluid that fills all the apparently empty space of the heavens,[3] and [2] "the matter of the heaven, in which the Planets are situated, unceasingly revolves, like a vortex having the Sun as its center" (Descartes 1984, 96; III.30). Descartes attributes different parts of the heavenly ether different speeds, namely, "those of its parts which are close to the Sun move more quickly than those further away" (*ibid*). But for all planetary bodies, the general explanation of their observed motion is the same. They move along their observed paths because they are carried by swirling pools of ether that circle the sun, much as straw that is placed in a swirling river is observed to move around the center of a vortex of water (*ibid*).

Newton sets out to refute Descartes's explanation of planetary motion in the concluding section of Book 2 of the *Principia* (entitled "Section 9: The circular motion of fluids"). His attempted refutation rests on two key assumptions: [1] Johannes Kepler's three laws of planetary motion accurately capture the observed orbital motion of planets around the sun (and of satellites around their primary planets),[4] and [2] any vortex that carries along a planet (or a satellite) must move in precisely the same way as the body it carries. From these assumptions, it follows that the heavenly ether of Descartes's model must move according to Kepler's laws, and, for Newton, this is precisely the problem. Based on

---

2 All references to the *Principles* are to the translation produced by V. R. Miller and R. P. Miller in Descartes (1984). All citations to the main body of the text are accompanied by the relevant part and section number.

3 For Descartes a vacuum in nature is impossible. Cf. Descartes (1984), 46–47; II.16.

4 Kepler published the first two of his laws in 1609 and the third in 1618. In simplified form, the three laws are as follows. Law 1 (The Ellipse Law): All of the planets orbit the sun along an ellipse that has the sun at one of the foci. Law 2 (The Area Law): A line segment drawn from the center of any planet to the center of the sun will sweep out equal areas in equal times. Law 3 (The Harmonic Law): The squares of the periods of any two planets is directly proportional to the cubes of their mean distances from the sun. For discussion of whether Newton accepted Kepler's laws as empirical laws, see Wilson (1974).

his analysis of fluid motion, he claims that there is no reasonable way to describe the thickness or the resistance of the ethereal fluid such that it would obey Kepler's Harmonic Law (Newton 1999, 787–788). He also determines that it is impossible for Descartes's "corporeal vortices" to move according to Kepler's Area Law (*ibid*, 789–790).[5] This is why Newton claims at the end of Book 2 that "the hypothesis of vortices can in no way be reconciled with astronomical phenomena." He has allegedly shown that a fluid ether cannot move according to the patterns that Kepler had identified in the observed motions of the planets. And what Newton aims to provide in Book 3 is an alternative and better explanation of planetary motions – one that accommodates what Kepler's laws express, one that assumes celestial bodies move in "free spaces without vortices," and one that has as its central feature a notion of gravity that is distinctively non-Cartesian.

Descartes's account of gravity is wedded to his Vortex Hypothesis and presented in Part IV of the *Principles* (entitled "Of the Earth"). He characterizes gravity as the "force of weight" that belongs to bodies on the earth and, specifically, as the feature of terrestrial bodies that explains why these bodies are heavy, that is, why they have a tendency to fall toward the center of the earth. Descartes's basic claim is that bodies exhibit this tendency to fall because of the heavenly ether. In his hypothetical system of the world, "the globules of this heavenly matter" have a natural and equal tendency to recede from the earth, and as they move toward the heavens, "they press down and drive below themselves some terrestrial parts into whose places they rise" (Descartes 1984, 191; IV.23). In turn, the precise weight of a terrestrial body (that is, the measure of its heaviness, or gravity) depends on the degree to which the body is impacted by the particles of ether that surround it. And according to Descartes, this degree of impact does not depend on "the quantity of matter in each body"; it depends on the specific type of matter of which the body is composed, that is, on the body's form (*ibid*, 192; IV.25).[6] Descartes clarifies by considering

---

5 Newton proves these two claims in the scholia to Book 2, Propositions 52 and 53, respectively. However, as first pointed out by George G. Stokes in a paper of 1845, the proofs in these scholia are based on some mistaken calculations. For discussion of Newton's calculation errors, see Cohen (1999), Section 7.9, and Smith (2005). For discussion of the specific assumptions about fluid resistance on which Newton's refutation of Descartes's Vortex Hypothesis rests, see Biener and Domski (forthcoming) and the works of George E. Smith that are cited therein.

6 Newton challenges Descartes's claim that a body's weight depends on a body's form in Corollary 1 of Proposition 6 of Book 3 of the *Principia*. I discuss Newton's argument in Section 2.2.

a piece of gold that weighs 20 times more than an equal volume of water. On his account, the water is less heavy, that is, has a lower weight, because, as a fluid body, it is composed of particles that move more quickly than the particles of gold, and, thus, in comparison to the gold, the water's particles are impacted to a lesser degree by the globules of ether that surround it (*ibid*; IV.25).[7]

Turning to Book 3 of the *Principia*, Newton makes no reference to an ether in his treatment of gravity, and he characterizes gravity as a "force of attraction" and an "attractive force," not as a "force of weight." However, his notion of gravity is not completely unlike Descartes's. For both of them, the gravity of a terrestrial body is broadly understood in terms of heaviness, and in general, it is a property, or quality, that is characterized in terms of the body's tendency to move, or fall, toward some other body. In somewhat different terms, for both of them, gravity explains why some bodies have a measurable weight, and it explains why some moving bodies are observed to deviate from straight-line inertial paths.[8] But here the agreement between Descartes and Newton seems to end.

For Newton, gravity is not exclusive to terrestrial bodies. He argues that it is a force that can be attributed to all natural bodies and to all the sensible and insensible parts of all natural bodies. He also demonstrates that this universal force can explain a variety of non-inertial motions. Gravity explains why terrestrial bodies are heavy bodies that have a tendency to fall toward the earth, but, according to Newton, gravity also explains why bodies in the heavens are observed to maintain curvilinear orbits around other heavenly bodies. Additionally, for Newton, the measure of gravity does depend on a body's mass. In the terms of Proposition 7 of Book 3 (III.7), "Gravity exists in all bodies universally and is proportional to the quantity of matter in each" (Newton 1999, 810). He also maintains that the measure of gravity depends on distance, and specifically, that its measure varies inversely as the square of the distance between two bodies. And nowhere in his

---

7 In the 1660s, Christiaan Huygens developed a mathematical version of the Vortex Hypothesis and a concomitant theory of gravity more robust than Descartes offers in the *Principles*. However, it was not until the 1690 publication of his *Discourse on the Cause of Gravity* (*Discours de la cause de la pesanteur*) that Huygens's version of the Vortex Hypothesis appeared in print. See Snelders (1989) for discussion of the relationship between the accounts of gravity forwarded by Huygens, Descartes, and Newton.

8 For a classic discussion of the similarities and differences between Descartes's and Newton's principles of inertia, see Chapter III ("Newton and Descartes") of Koyré (1965).

treatment of gravity does Newton assign a positive role to the factors that play centrally in Descartes's account. Neither the type of matter of which a body is composed nor the impacts that a body suffers from the medium in which it is situated inform the notion of universal gravity that Newton defends.

This points to an additional difference between Descartes and Newton: They are committed to competing conceptions of what it means to explain gravity. Descartes provides an answer to the question of why there is a "force of weight" among terrestrial bodies, that is, of why they exhibit a tendency to fall to earth. It's because they are being pushed downward by the ether. In contrast, Newton does not specify a cause for either terrestrial or celestial gravity. One reason that he does not, he tells us, is because his focus in the *Principia* is on the "quantities and mathematical proportions" of attractive and impulsive forces that are found in nature (*ibid*, 588). As a consequence, the question of what natural circumstances are causally responsible for these natural forces – of what underlying natural events are producing the sorts of forces that he is studying – is not one with which he will engage. The result is that the task of explaining gravity takes on a non-Cartesian form. As indicated by what Newton accomplishes in Book 3, explaining gravity involves determining its scope, that is, identifying the bodies to which it belongs, and providing a robust mathematical characterization of its measure, that is, identifying the laws that it obeys. But it does not require identifying the reasons that this force is in nature and belongs to natural bodies. It also does not require explaining why gravity obeys one set of laws rather than some other. For Newton, these are not questions that he must address, because he is not "considering in this treatise ... the species of forces and their physical qualities" (*ibid*).[9]

---

9 The remarks I cite in this paragraph come from the Scholium to I.69, where Newton writes:

> I use the word "attraction" here in a general sense for any endeavor whatever of bodies to approach one another, whether that endeavor occurs as a result of the action of the bodies either drawn toward one another or acting on one another by means of spirits emitted or whether it arises from the action of aether or of air or of any medium whatsoever – whether corporeal or incorporeal – in any way impelling toward one another the bodies floating therein. I use the word "impulse" in the same general sense, considering in this treatise not the species of forces and their physical qualities but their quantities and mathematical proportions, as I have explained in the definitions.
>
> (Newton 1999, 588)

Newton elaborates on this difference with Descartes in the General Scholium, a nine-paragraph concluding section that was added to the second edition *Principia* and maintained in the third edition with minor revisions. He had already explained in the first edition that he did not investigate the causes of the attractive and impulsive forces that he was examining because of the mathematical focus of his investigations. In the General Scholium, he offers an additional explanation and clarifies that he has refrained from identifying the causes of these forces, and the cause of gravity in particular, because of the distinctive kind of natural philosophy that he has pursued in the *Principia*. His natural philosophy is "experimental philosophy," he reports, and it is one that does not allow for the use of "hypotheses."[10]

Descartes's Vortex Hypothesis is clearly one of the "hypotheses" that Newton excludes from his experimental philosophy. In all editions of the *Principia*, he ends Book 2 with his attempted refutation of the "hypothesis of vortices," and in both the second and third editions, he opens the General Scholium with the assertion that "The hypothesis of vortices is beset with many difficulties" (*ibid*, 939). But

---

The explanation to which Newton is referring is included at the end of Definition 8:

> I use interchangeably and indiscriminately words signifying attraction, impulse, or any sort of propensity toward a center, considering these forces not from a physical but only from a mathematical point of view. Therefore, let the reader beware of thinking that by words of this kind I am anywhere defining a species or mode of action or a physical cause or reason, or that I am attributing forces in a true and physical sense to centers (which are mathematical points) if I happen to say that centers attract or that centers have forces.

> (*ibid*, 408)

10 Shapiro (2004) notes that Newton first uses the term "experimental philosophy" to describe his program of natural philosophy in a draft of Query 23 of the *Opticks* from 1706 (Shapiro 2004, 189). However, the first time Newton uses the term in print is in the General Scholium of the second edition (1713) *Principia*. In what follows, I focus on the relationship between Newton's characterization of experimental philosophy in the General Scholium and his argument for universal gravity and leave aside the question of whether Newton's practice of natural philosophy in general meets the standards of the experimental philosophy that he there describes. Consequently, I do not engage the long-standing question of whether Newton consistently avoids using the sorts of "hypotheses" that he publicly rejects. Now-classic treatments of this issue can be found in Koyré (1965) and Cohen (1966), and, in regard to Newton's optical works, in Shapiro (1993). See Wilson (2019) and Biener and Domski (forthcoming) for more recent contributions to this discussion, and see Shapiro (2004) for a trenchant account of the various published and unpublished writings in which Newton uses the term "experimental philosophy" to distance his natural philosophy from the type of hypothetical natural philosophy that he associates with Descartes and Leibniz.

in the penultimate paragraph of the General Scholium, Newton is making a broader point about what is permissible in his program of natural philosophy. He claims that "whatever is not deduced from the phenomena must be called a hypothesis" and asserts that no hypotheses whatsoever – no matter whether they are "metaphysical or physical, or based on occult qualities, or mechanical" – have any "place in experimental philosophy. In this philosophy," he continues,

> propositions are deduced from the phenomena and are made general by induction. The impenetrability, mobility, and impetus of bodies, and the laws of motion and the law of gravity have been found by this method. And it is enough that gravity really exists and acts according to the laws that we have set forth and is sufficient to explain all the motions of the heavenly bodies and of our sea.
>
> (*ibid*, 943)[11]

According to Newton, he has shown in the *Principia* that there is a force of gravity that acts according to specific laws, namely, that it is a force with a measure that varies directly with mass and varies inversely as the square of the distance between two bodies. He has also shown that this force can be used to explain "the phenomena of the heavens and of our sea" (*ibid*). What he has not done is offered any explanation for why "gravity really exists" and acts according to the laws that have been identified. There are two reasons that he hasn't. He has "not as yet been able to deduce from phenomena" any such explanation for gravity, and, additionally, he refuses to "feign hypotheses" – or, in his more familiar terms, "Hypotheses non fingo" (*ibid*). In brief, Newton has refrained from assigning a cause to gravity, because he has pursued in the *Principia* a program of experimental philosophy, not a program of hypothetical philosophy.[12]

11 My translation of these remarks is slightly different than the one in Newton (1999). Cohen and Whitman use "In this experimental philosophy" as the translation of "In hac Philosophia," whereas I use the literal "In this philosophy." The addition of "experimental" in the Cohen and Whitman translation was previously pointed out by Alan Shapiro (cf. Shapiro 2004, 187, Note 2).

12 Newton presented these remarks about the differences between experimental philosophy and hypothetical philosophy as a means of replying to critics of the first edition *Principia* who alleged that he left his account of gravity open to objection precisely because he did not offer a causal explanation for gravity. For instance, in his *Discourse on the Cause of Gravity* (1690), Huygens claims that Newton's version of gravitational attraction should be rejected, because its features cannot be

Considering how Descartes presents and defends his Vortex Hypothesis in Part III of the *Principles*, the differences that Newton invites us to see between experimental philosophy and hypothetical philosophy are not unfounded. For instance, Descartes gives no indication that his explanatory model has been "deduced from the phenomena" in any meaningful way. He does claim at the opening of Part III that we should begin our inquiries into the visible world by describing empirical events. However, the reason we do this, Descartes says, is only so that we can isolate the "principal natural phenomena" that we wish to explain, "not in order that we may use them [the phenomena] to prove anything" (Descartes 1984, 85; III.4; bracketed phrase added). Additionally, as Descartes defends his Vortex Hypothesis, he warns us not to think of his explanatory model as providing a picture of what really exists in the natural world. As he puts it,

> I shall set forth here the hypothesis which seems to me the simplest and most useful of all; both for understanding the phenomena and for enquiring into their natural causes. And yet I give warning that I do not intend it to be accepted as entirely in conformity with the truth, but only as an hypothesis {or supposition which may be false}.
>
> (*ibid*, 91; III.19)[13]

Descartes's model of the heavens is premised on the existence of a pervasive material ether with particles moving at varying speeds. But, according to Descartes, accepting the model does not require that we accept that such an ether actually exists. To accept the model is to use

---

explained in terms of the principles of mechanics or the rules of collision (cf. Cohen 1999, Section 6.11, 153). The criticisms expressed by G. W. Leibniz follow along broadly similar lines. As Huygens, Leibniz maintains that Newton's gravity cannot be explained in terms of mechanical causes. He makes the additional point that Newton's gravity must, as a consequence, be understood as some sort of "occult quality" of bodies – as a primitive force that resides in material objects and acts on other bodies in an unintelligible way. For Leibniz, it is on these grounds that Newton's notion of gravity should be rejected. Leibniz forwards different versions of this criticism in writings such as "Against Barbaric Physics" (1710–1716?) and the fourth and fifth letters he wrote to Samuel Clarke (cf. Sec. 45–46 of Letter Four and Sec. 112–115 of Letter Five), all of which are available in Leibniz (1989). See also Leibniz's letter to Hartsoeker from 10 February 1711, which I briefly discuss in Section 4.2. That letter and Hartsoeker's reply from 13 March 1711 are published in English translation (from their original French) in Leibniz (1712).

13 Following Miller and Miller's convention in Descartes (1984), squiggly brackets are used to isolate text that was added to the 1647 French edition of the *Principles*.

it as a calculating device, one that allows us to predict the motions and positions of heavenly bodies and thereby save the phenomena.

However, deciding to accept one hypothesis from among several is not, for Descartes, merely a matter of determining which hypothesis allows for more accurate predictions; we must also consider whether a hypothesis is consistent with what is known to be true of nature.[14] Descartes clarifies in the Author's Preface to the French edition (1647) of the *Principles* that all our knowledge of nature derives from three foundational principles. There are two principles "concerning immaterial and Metaphysical things," namely, that the soul exists and that God exists both as the "author of everything which is in the world" and as "the source of all truth" (Descartes 1984, xxii). There is also one principle of "corporeal or physical things," namely, "that there are bodies extended in length, width, and depth, which have diverse figures and are moved in diverse ways" (*ibid*). According to Descartes, these three principles are among "the most evident and most clear [things] which the human mind can know" (*ibid*, xxi), and they are such that their certainty and clarity become manifest when we correctly reason about our innate ideas.[15] The three foundational principles are also "such that one can deduce from them the knowledge of all other things which are in this world" (*ibid*, xxii). Descartes demonstrates this point in Part II of the *Principles* (entitled "Of the Principles of Material Objects") by showing that central truths about the physical world – for instance, that all material objects are governed by three laws of nature (cf. *ibid*, 57–62; II.36–42) – are derivable from our understanding of God's nature and from our awareness that bodies are essentially extended, that is, from our knowledge that "the nature of matter, or of body considered in general" consists "only in the fact that it is a thing possessing extension in length, breadth, and depth" (*ibid*, 40; II.4).

In the General Scholium, Newton voices opposition to this feature of Descartes's natural philosophy as well. The relevant remarks are presented in the paragraph that precedes his explanation of why he

14 It is on these grounds that Descartes argues that his Vortex Hypothesis is superior to the explanations of celestial motions that had been forwarded by Ptolemy, Tycho, and Copernicus (cf. *ibid*, 89–91; III.16–19). See Domski (2019) and McMullin (2008, 2009) for further discussion.
15 For instance, in Part II Descartes notes that "the perceptions of our senses do not teach what really exists in things," and he instructs us to concentrate on "those ideas which nature endowed" our understanding so that we can perceive that "the nature of body" consists "in extension alone" (Descartes 1984, 40; II.3–4). In Section 4.3 I address the connection between Descartes's characterization of body and his account of the explanations that should be used in natural philosophy.

has not identified a cause of gravity. There Newton discusses the differences between God and humans and notes that God is a "most wise" being who "senses and understands all things" – a being who is "all force of sensing, of understanding, and of acting, but in a way not at all human, in a way not at all corporeal, in a way utterly unknown to us" (Newton 1999, 942). According to Newton, our limited human capacities for sensing and understanding prevent us from knowing how God operates. They also prevent us from having the kind of knowledge that is reserved for the divine. We "certainly do not know what is the substance of any thing," he says.

> We see only the shapes and colors of bodies, we hear only their sounds, we touch only their external surfaces, we smell only their odors, and we taste their flavors. But there is no direct sense and there are no indirect reflected actions by which we know inner-most substances; much less do we have an idea of the substance of God.
>
> (*ibid*)

For Newton, there is no option of relying exclusively on our capacity to reason to discover what is true about material bodies, let alone to gain knowledge of which properties are essential to these bodies. To gain any understanding of existing things, we have to rely on our senses, and these, he tells us, are not equipped to provide us knowledge of the "innermost substances" that belong to anything that exists. Consequently, and contra Descartes, awareness of the true and essential natures of God and of bodies simply cannot direct us in our investigations into the natural world.

The second and third editions of the *Principia* leave little question about how far Newton's opposition to Descartes's natural philosophy extends. In the newly added General Scholium, Newton rejects the use of hypotheses in natural philosophy. He also rejects the possibility that we can know "the substance of any thing," and thus that we can gain the knowledge of God and of bodies that Descartes makes foundational to his natural philosophy. Newton also portrays his experimental philosophy as a non-Cartesian way of investigating nature that makes use of empirical evidence in a different and more robust way than Descartes allows. How exactly Newton uses empirical evidence to establish that "gravity really exists" and "exists in all bodies universally" is a significant question that lingers.

1.2 Rule 3 and the Plan of This Book

In the *Principia*, Newton explicitly associates the method of experimental philosophy with the method by which he establishes the existence of a universal force of gravity just once, in the penultimate paragraph of the General Scholium. As we saw above, there he says rather generally that the method of experimental philosophy involves using phenomena and induction to demonstrate claims about the natural world – in "this philosophy propositions are deduced from the phenomena and are made general by induction," he tells us (*ibid*, 943). And he reports that it was by means of this method that he found that bodies have the properties of impenetrability, mobility, and inertia, and that they obey the laws of motion and the law of gravity (*ibid*).

The other explicit statements that Newton makes in the *Principia* about the methodology that he uses to establish the existence of a universal force of gravity are cast in broader terms and focused on the relationship between what is achieved in Books 1 and 2 and what is achieved in Book 3. In these instances, Newton stresses in particular that his project of explaining the system of the world in Book 3 depends on the general and mathematical propositions concerning forces and motions that are demonstrated in the opening books. For instance, in the Author's Preface to the first edition, he explains that

> the basic problem [lit. whole difficulty] of philosophy seems to be to discover the forces of nature from the phenomena of motions and then to demonstrate the other phenomena from these forces. It is to these ends that the general propositions in books 1 and 2 are directed, while in book 3 our explanation of the system of the world illustrates these propositions. For in book 3, by means of propositions demonstrated mathematically in books 1 and 2, we derive from celestial phenomena the gravitational forces by which bodies tend toward the sun and toward the individual planets. Then the motions of the planets, the comets, the moon, and the sea are deduced from these forces by propositions that are also mathematical.
>
> (*ibid*, 382)[16]

---

16 For discussion of how Newton's attitude toward mathematical knowledge and the methods used in geometry inform his remarks in the Author's Preface, see Guicciardini (2009, Chapter 13), and Domski (2003).

The mathematical character of the propositions that are proven in Books 1 and 2 is also stressed in the prefatory paragraph that opens Book 3. Again using the term "philosophy" to refer to natural philosophy, Newton notes that the natural philosophy of the final book is based on the "strictly mathematical" principles presented in the first two books:

> In the preceding books I have presented principles of philosophy that are not, however, philosophical but strictly mathematical – that is, those on which the study of philosophy can be based. These principles are the laws and conditions of motions and of forces, which especially relate to philosophy ... It still remains for us to exhibit the system of the world from these same principles.
>
> (*ibid*, 793)[17]

As Newton presents it, the *Principia* as a whole tracks two stages. The first is mathematical and included in Books 1 and 2; the second is natural philosophical and included in Book 3. And Newton's common refrain is that the natural philosophical project of exhibiting the system of the world, and explaining the motions of celestial bodies in terms of gravity, depends both on empirical evidence and on the mathematical claims already proven.

Newton briefly elaborates on the sense in which the natural philosophy of Book 3 is based on the propositions of Books 1 and 2 in the Scholium to I.69. There he addresses how a natural philosophy that starts from mathematical propositions should proceed and indicates that in Book 3 he has come "down to physics" and "compared with the phenomena" the "quantities and mathematical proportions" that were established in Books 1 and 2 (*ibid*, 588–589). He additionally claims that it was by making this comparison that he has "found out which conditions [or laws] of forces apply to each kind of attracting bodies" (*ibid*, 589). By this account, then, it was by comparing mathematical propositions with empirical evidence that Newton was able to determine the laws that gravity obeys and to identify the bodies to which gravity belongs.[18]

---

17 Given the central role of the laws of motion in Books 1 and 2 – laws which Newton reports at the opening of the *Principia* are ones that capture the motions of natural bodies (*ibid*, 416–417) – it is not entirely clear how to interpret Newton's claim that the propositions that are demonstrated in the opening books are "strictly mathematical." For discussion of this issue, see Biener and Schliesser (2017), Schliesser (2013), and the related works cited therein.

18 The remarks I quote here are presented immediately after Newton reports that he will focus in the *Principia* on the "quantities and mathematical proportions" of

Taking a closer look at what Newton does in Book 3, there is a specific progression that characterizes the "comparisons" that he makes between the phenomena and what's been mathematically demonstrated in Books 1 and 2. Namely, in proving several of the propositions that contribute to the argument for universal gravity, Newton reasons from an experimentally supported explanation of the motions of *terrestrial* bodies to an explanation of the motions of *celestial* bodies. There are, of course, a number of sophisticated moves that Newton makes to go from terrestrial gravity to universal gravity – moves involving empirical evidence and mathematical propositions from Books 1 and 2, and also, at times, some unstated assumptions. But at critical moments of the argument Newton reasons from terrestrial gravity to celestial gravity by extending to celestial bodies results that have been gathered from experiments that had been conducted on terrestrial bodies. For instance, in the proof for III.4, Newton uses the results of pendulum experiments that had been conducted by Christiaan Huygens to establish that a single inverse-square force of gravity governs the fall of terrestrial bodies. Newton then extends this result to the moon and argues that a single inverse-square force of gravity also explains why the moon maintains its orbit around the earth. This is not the only place in Book 3 where Newton progresses from a claim about bodies on which experiments *have been made* to a claim about a body on which experiments *cannot been made*. As we will see in Sections 2.1 and 2.2, using experimental results to generate explanations of why celestial bodies are observed to move as they do is a hallmark of the strategy that Newton uses to prove several other propositions that contribute to the argument for universal gravity.

---

attractive and impulsive forces that are found in nature and not investigate their physical qualities or causes (see Note 9). In the paragraph that follows, Newton writes:

> Mathematics requires an investigation of those quantities of forces and their proportions that follow from any conditions that may be supposed. Then, coming down to physics, these proportions must be compared with the phenomena, so that it may be found out which conditions [or laws] of forces apply to each kind of attracting bodies. And then, finally, it will be possible to argue more securely concerning the physical species, physical causes, and physical proportions of these forces.

(Newton 1999, 588–589)

See Smith (2016) for a reading of the method that Newton adopts in the *Principia* that takes the Scholium to I.69 as its point of departure.

This is also a type of reasoning that Newton explicitly endorses in the section that opens Book 3 in the second and third editions of the *Principia*. The section is entitled "Rules for the Study of Natural Philosophy" (*Regulae Philosophandi*), and in it, Newton sets forth directives concerning the inferences that the experimental philosopher ought to draw when particular kinds of empirical evidence are available.[19] Come the third edition of 1726, the four Rules are stated as follows:

Rule 1: No more causes of natural things should be admitted than are both true and sufficient to explain their phenomena.

Rule 2: Therefore, the causes assigned to natural effects of the same kind must be, so far as possible, the same.

Rule 3: Those qualities of bodies that cannot be intended and remitted and that belong to all bodies on which experiments can be made should be taken as qualities of all bodies universally.

Rule 4: In experimental philosophy, propositions gathered from phenomena by induction should be considered either exactly or very nearly true notwithstanding any contrary hypotheses, until yet other phenomena make such propositions either more exact or liable to exceptions. (*ibid*, 794–796)

As the Rules are stated in the third edition, there is a consistent emphasis on what the experimental philosopher is licensed to infer when a particular set of evidentiary circumstances obtain. Newton identifies in Rules 1 and 2 how many and which sort of natural causes

19 In the first edition of the *Principia*, the opening section of Book 3 was entitled "Hypotheses" and included nine statements, each of which was labeled "Hypothesis." In the second edition, Hypothesis I and Hypothesis II were slightly modified and presented as Rule 1 and Rule 2, respectively. The only difference is that, in the second edition, Newton includes a sentence in the commentary of Rule 1 that did not appear in the commentary of Hypothesis I (cf. Note 52). As for the other Hypotheses of the first edition, Hypothesis III is not included in the later editions; Hypothesis IV is renamed Hypothesis 1 and presented in the main part of Book 3, between Propositions 10 and 11; and the remaining five Hypotheses are listed in the "Phenomena," the section of Book 3 that immediately follows the "Rules for the Study of Natural Philosophy." For more on how the Hypotheses of the first edition compare to the Rules of the later editions, see especially Newton (1999, 794, Note aa), Cohen (1966), Cohen (1971, 23–26), Biener (forthcoming, Section 1.3), and Koyré (1965). For discussion of why Newton replaced Hypothesis III with Rule 3 in the second and third editions, see the works referenced in Note 35.

"should be **admitted**" and "**assigned**"; in Rule 3 he tells us which qualities "should be **taken** as qualities of all bodies universally"; and in Rule 4 he describes how "propositions gathered from phenomena by induction should be **considered**."[20] There is no indication that, under particular evidentiary circumstances, an experimental philosopher could justify a claim about how nature actually and in fact operates. Fitting of Newton's pronouncements in the General Scholium that we cannot expect the "innermost substance" of things to be disclosed to us through our senses and that God acts "in a way utterly unknown to us," with the Rules Newton leaves open the possibility that what can be legitimately inferred about natural causes and the qualities of natural bodies from empirical evidence may not capture the real and divinely created order of things.[21]

At various times in Book 3, Newton explicitly refers to one or more of the Rules to justify a key inference in his argument for universal gravity. He makes three references to Rule 1, four references to Rule 2, and one reference to Rule 3 and to Rule 4. (The specific places in Book 3 where the Rules are referenced are discussed in Sections 2.1 and 2.2.) But in regard to the type of reasoning that is characteristic of Book 3 – reasoning that progresses from claims about particular experimental cases to claims about bodies on which experiments cannot

---

20 Throughout what follows I use boldfaced font to indicate where I have added emphasis to texts that are directly quoted from Newton. Following Cohen and Whitman, the portions of the direct quotes that were emphasized by Newton are presented in italics.

21 That Newton's intention in the third edition is to identify the conclusions that empirical evidence warrants the experimental philosopher to accept, rather than to identify what empirical evidence reveals about the true workings of nature, is also signaled by the changes he made to various versions of the Rules prior to 1726. In earlier drafts, the language he uses indicates that the experimental philosopher could make reliable inferences about the actual makeup of nature on the basis of empirical evidence. For instance, in an annotated copy of the first edition *Principia* that Newton had supplied to John Locke, the draft of what would become Rule 3 of the second and third editions was labeled "Hypothesis III" and read: "Hypoth. III. The qualities of bodies that cannot be intended and remitted, and that belong to all bodies in which one can set up experiments, **are the qualities of bodies universally** [*sunt qualitates corporum universorum*]" (cf. Cohen 1971, 24; boldface added). Newton here asserts that the types of qualities identified are those that, in actuality, belong to all natural bodies. In the published version of Rule 3, in contrast, Newton uses the phrase "should **be taken** as qualities of all bodies universally." There is a similar revision to Rule 2 between the second and third editions of the *Principia*. In the second edition, Rule 2 states that "for natural effects of the same kind the causes **are the** same" (cf. Cohen 1971, 262; boldface added), whereas in the third edition, Newton makes a more tempered claim and says that the same causes ought to be "**assigned to**" the same effects.

be made – the directive communicated by Rule 3 is especially significant. Bracketing for a moment its finer details, the Rule indicates that when some specified kind of empirical evidence has been gathered, it is legitimate to accept that a quality of bodies belongs to "all bodies universally." In the commentary that follows the Rule, Newton clarifies that the set of "all bodies universally" includes all the natural bodies that we have yet to observe and also all the natural bodies that are "beyond the range of [our] senses" (*ibid*, 795; bracketed term added). Accordingly, Rule 3 communicates that there are specific evidentiary circumstances under which the experimental philosopher is justified in drawing inferences from what *is* perceived to what *cannot* be perceived. Putting the point differently, the Rule tells us that there are evidentiary circumstances under which it is legitimate to reason from claims that have been confirmed by empirical evidence to claims for which no direct empirical evidence can be gathered. Rule 3 thus supports the possibility of extending results that have gathered from bodies on which experiments have been made to bodies on which experiments cannot be made. And in this respect, the Rule is an endorsement of a type of reasoning that is critical to Newton's argument for universal gravity in Book 3.[22]

Because of this link between Rule 3 and the argument strategy that Newton uses in Book 3, several commentators have come to view the Rule as "a primary statement of Newton's philosophy of science" (Cohen 1971, 24).[23] Commentators have also shown that a closer examination of the connection between the Rule and the account of gravity that Newton presents in Book 3 leaves us with significant questions

---

22 Due to the wide scope of the bodies that Rule 3 covers, the Rule has long been characterized as Newton's "inductive principle," or rule for generalization. See, for instance, the remarks of Henry Pemberton and William Whewell that are discussed in Biener (forthcoming, Section 1.2). Because Newton indicates that Rule 3 allows us to make claims about the insensible parts of bodies based on what is sensed, more recent scholars have characterized Rule 3 as Newton's "principle of transduction." See, for instance, Mandelbaum (1964), McGuire (1968), and Okruhlik (1989). See also Belkind (2017), who helpfully groups several recent approaches to Rule 3 according to the type of reasoning – whether inductive or transductive – that commentators claim the Rule is ultimately meant to support. For reasons I set forth immediately below, I do not read Rule 3 as a statement that either type of reasoning is more fundamental than the other, and thus, I continue to refer to it as a "universalizing rule" rather than a rule for induction or for transduction.

23 This sentiment can also be found in Cohen (1966), Mandelbaum (1964), McGuire (1968), and Okruhlik (1989).

about gravity's features and about the consistency of Newton's claims regarding our knowledge of bodies.

Consider, for instance, Newton's claim that, once appropriate empirical evidence has been collected, gravity can be universalized on the basis of Rule 3 (Newton 1999, 796). The question we face here is why the Rule can be applied to gravity at all, and the question arises because of Newton's use of "intended and remitted" (*intendi & remitti*) in the Rule's opening clause. Recall what Rule 3 states: "Those qualities of bodies that cannot be intended and remitted [*intendi & remitti*] and that belong to all bodies on which experiments can be made should be taken as qualities of all bodies universally" (*ibid*, 795; bracketed Latin terms added). It has long been accepted that Newton's use of *intendi & remitti* should be read along the lines of the medieval theory of the latitude of forms, according to which a quality can be "intended and remitted" if and only if it is a quality that can vary in degree, that is, that can increase or decrease, or be "augmented and diminished" (Cohen 1971, 24, Note 6).[24] Consequently, if Rule 3 can be applied to gravity, as Newton says it can, then it seems that as per the Rule's opening clause, gravity must be a "quality of bodies that **cannot** be intended and remitted," that is, it must be a quality with a measure that cannot be increased or decreased. However, the gravity of one and the same body is "remitted" – its measure does decrease – as the body grows more distant from the body to which it is attracted. Indeed, five lines after claiming that Rule 3 can be applied to gravity, Newton emphasizes that as "bodies recede from the earth," their "[g]ravity is **diminished**" (*ibid*, 796; boldface added). This puzzle about gravity's features – about whether, for Newton, gravity is or is not a

---

24 It is because they take Newton to be adopting the medieval use of the phrase *intendi & remitti* that, in Newton (1999), Cohen and Whitman include in brackets "i.e., qualities that cannot be increased and diminished" immediately after their literal translation of "Qualitates corporum quæ intendi & remitti nequeunt" (Newton 1999, 795). Additional support for reading the "cannot be intended and remitted" of Rule 3 as "cannot be increased or diminished," or as "cannot be augmented and diminished," is offered by Cohen (1999) and McGuire (1968), who highlight the connection between Newton's use of the phrase with its use by English contemporaries such as John Keill, John Harris, John Clarke, and Henry Pemberton (cf. Cohen 1999, 200; McGuire 1968, 241–244). Cohen also points out that an appreciation for Newton's appropriation of the medieval theory of the latitude of forms seems to be why, in her modernized French translation of the *Principia* of 1759, Émilie du Châtelet opted to translate "Qualitates corporum quæ intendi & remitti nequeunt" as "'les qualités des corps qui ne sont susceptibles ni d'augmentation ni de diminution' (Paris, 1759, vol. 2, p. 3)" (Cohen 1971, 24–25, Note 6).

quality that can be intended and remitted – has played centrally in the commentaries of scholars such as Okruhlik (1989), Janiak (2008), and Ducheyne (2012).

Scholars such as McGuire (1970) and McMullin (1978) have helped illuminate a different puzzle surrounding the connection between Rule 3 and the notion of gravity that Newton defends in Book 3. For them, the central interpretive issue turns on remarks that Newton makes in the commentary that accompanies Rule 3, where he explains why the Rule should be accepted. What is of particular interest are Newton's claims about the qualities of extension, hardness, impenetrability, inertia, and mobility. He notes that these qualities of bodies are "known to us only through our senses," and that they are qualities that we infer belong to "bodies beyond the range of these senses" (*ibid*, 795). He then appears to claim that these sorts of inferences, which proceed from whole bodies to the parts of bodies, and from what is perceived to what is imperceptible, are legitimate inferences to draw because of the atomistic structure of bodies. He writes:

> The extension, hardness, impenetrability, mobility, and force of inertia of the whole arise from [*oritur*] the extension, hardness, impenetrability, mobility, and force of inertia of each of the parts; and thus we conclude that every one of the least parts of all bodies is extended, hard, impenetrable, movable, and endowed with a force of inertia. And this is the foundation of all natural philosophy.
>
> (*ibid*, 795–796; bracketed term added)

In the opening clause of this passage Newton appears to be forwarding the basic tenet of atomism, namely, that sensible bodies are composed of insensible parts, and also the related tenet that the qualities of a sensible body originate from the qualities of the body's insensible parts. This suggests that, for Newton, drawing inferences from what is evident to our senses to claims about the qualities that belong to all bodies universally – including those bodies that are insensible – requires accepting that nature is ordered in the atomistic way that he describes. The further suggestion is that, for Newton, any reasoning that takes us from bodies on which experiments have been performed to bodies on which experiments either have not or cannot be performed also rests on a commitment to an atomist account of reality. Consequently, despite Newton's claim in the General Scholium that neither God's activity nor the real, "innermost substances" of bodies can be known to us, the commentary that accompanies Rule 3 leaves us with a picture of the Rule, and also of Book 3, according to which pursuing experimental philosophy appears to depend on our knowledge that God imposed on natural bodies an atomistic structure and ordering.

In what follows, I provide a new and unexplored way of addressing the interpretive problems surrounding Newton's justification for Rule 3, and also of addressing the general problems surrounding the Rule's connection to gravity. My approach to these problems has remained unexplored, because at its heart is a new and unexplored way of reading Rule 3.

To cash this out, recall again what the Rule states:

> Those qualities of bodies that cannot be intended and remitted and that belong to all bodies on which experiments can be made should be taken as qualities of all bodies universally.
>
> (*ibid*, 795)

According to the long-standard interpretation, the Rule provides us a description of the features that belong to a single set of universalizable qualities, and it communicates the two necessary and sufficient conditions that are met by the members of that single set. For instance, in the *Guide* that precedes Newton (1999), Cohen says that the message of Rule 3

> is that there is a certain set of "qualities" that (1) are found in all bodies within the range of our direct experience on earth and (2) do not vary, and that these are to be considered qualities of all bodies universally, that is, of bodies everywhere in the universe.
>
> (Cohen 1999, 199)

Cohen's characterization helpfully underscores that, with Rule 3, Newton is not making a positive claim about the kinds or types of qualities that really and in fact belong to bodies everywhere in the universe. Rather, Newton is identifying the features of a set, or group, of qualities, which we can legitimately infer are universal. That is, he is describing a set of qualities that should be understood as belonging to the general class of universal qualities.

On this point, I fully agree with Cohen. With Rule 3, Newton is not providing the means for identifying qualities that are really and in fact universal. However, unlike Cohen, I claim that Rule 3 refers not to a single set of qualities but to two sets. The members of one set are qualities of bodies that "cannot be intended and remitted," and the members of the other set are qualities that "belong to all bodies on which experiments can be made." And, on this Two-Set Reading, the message of Rule 3 is that a quality can be considered a universal quality if it belongs *either* to one set *or* to the other. In other words, with Rule 3 Newton is describing two sufficient conditions for a quality of bodies to be universalized, not two necessary and sufficient conditions.

I grant that there is no definitive textual evidence that picks out the Two-Set Reading as the correct way of interpreting Rule 3. However, the same can be said of the One-Set Reading. Both interpretations allow us to connect a variety of claims from Book 3, but, arguably, neither is uniquely supported by Newton's remarks. My defense of the Two-Set Reading therefore takes on a particular form. I aim to show that this reading of Rule 3 is compatible with the same key texts from the *Principia* that can be reconciled with the One-Set Reading. I also aim to show that, in comparison to the One-Set Reading, the Two-Set Reading provides us a more straightforward way of understanding the Rule's connection to gravity and a more straightforward way of resolving various interpretive puzzles that stem from Newton's commentary on Rule 3. There is the puzzle about atomism that was already noted. There are also puzzles surrounding Newton's claim that the Rule can be applied to a case in which an experimental philosopher has gathered empirical evidence from a single experiment, as well as his claim that inertia can be considered an inherent force that is essential to bodies, whereas gravity cannot. I cash out these interpretative issues in Section 2.3.

In the opening sections of Chapter 2, I clarify the role of the four Rules for the Study of Natural Philosophy in Newton's argument for universal gravity. I begin in Section 2.1 with a brief overview of the various instances at which Rules 1, 2, and 4 are applied in Book 3, and then, in Section 2.2, I take a more careful look at Newton's use of Rule 3 in Corollary 2 of III.6. My aim in these two sections is to illuminate how Newton uses the Rules to extend to celestial bodies what has been empirically and experimentally confirmed of terrestrial bodies. I thus offer a focused reading of Newton's argument for universal gravity that highlights his use of empirical and experimental evidence to justify claims about bodies on which no experiments can be made.[25]

---

25 There are, of course, a number of other strategies one could adopt to illuminate the status of empirical evidence in Newton's argument for universal gravity, as made clear by the vast (and ever-growing) literature on the nuances of Newton's reasoning in Book 3. For recent discussions of Newton's argument for universal gravity, see the relevant chapters of Biener and Schliesser (2014), Iliffe and Smith (2016), Janiak and Schliesser (2012), and Schliesser and Smeenk (forthcoming), as well as the many related works cited therein.

The bulk of Chapter 3 is dedicated to defending the plausibility of the Two-Set Reading. I start by showing in Section 3.1 that the Two-Set Reading can accommodate the same key selections from the *Principia* that are compatible with the One-Set Reading. I then show in Section 3.2 that reading Rule 3 as identifying two sets is consistent with Newton's use of the Latin *quæ ... quæque* construction to express the Rule and also with his use of that construction elsewhere in the *Principia*. In Section 3.3, I apply the Two-Set Reading to Newton's brief justification for Rule 3 and to his two examples of cases to which the Rule can be applied. One example involves applying the Rule to gravity. The other involves applying it to the quality of divisibility. (It is in this instance that Newton applies the Rule to evidence that has been gathered from a single experiment.) Ultimately, I claim that the Two-Set Reading gives us a clearer view than the One-Set Reading of why Rule 3 can be applied to both of the cases that Newton presents.

In Chapter 4, I explore two additional advantages of adopting the Two-Set Reading of Rule 3. As a preliminary step, I consider in Section 4.1 the status of the atomist picture of nature that Newton uses to justify Rule 3. Here I connect his claims about reasoning from the qualities of the parts of bodies to the qualities of whole bodies with the rhetoric and general argument strategy that characterize his presentation of the Rules. As I read Newton, he is neither claiming that the atomist picture of reality is true nor trying to convince his readers that it is true. Rather, he is assuming that this is the picture of reality that his readers already accept and is showing that the reasoning that characterizes his experimental philosophy is consistent with the image of nature and natural bodies that they are bringing to the text. Building off of this reading, I argue in Section 4.2 that, on the Two-Set Reading of Rule 3, Newton circumvents the questions about his distinction between inertia and gravity that are prompted by the One-Set Reading, and additionally, that he has adequate grounds for claiming that inertia should be considered an inherent and essential force, whereas gravity should not. I conclude the book by showing that from the Two-Set Reading we gain a more crystallized view of the differences between Newton's experimental philosophy and Descartes's hypothetical philosophy. Namely, I argue in Section 4.3 that, based on the reading that I offer, we gain a firmer appreciation of how Newton is able to toe a middle line between speculation and truth that is unavailable to Descartes. Insofar as Rule 3 directs the experimental philosopher to universalize qualities based on empirical evidence, Newton can avoid the charge that gravity is a physical hypothesis. And insofar as Rule 3 allows the experimental philosopher to regard some qualities that have been universalized as the general but nonessential qualities of bodies, Newton can explain the system of the world in reference to a universal force of gravity that really exists without requiring knowledge of what is essential to natural bodies.

My broader aim in what follows is to provide an image of how Newton intended his program of experimental philosophy to be understood by readers of the final publicly available version of the *Principia*. It is for this reason that I give most sustained attention to the published remarks from the third edition of 1726. It is also why I generally set to the side questions about how the natural philosophy that Newton pursues in the third edition might be connected to his various other published and unpublished remarks about the reasoning and methods that are appropriate to the study of nature. These questions are important. They'll simply have to wait for another time.

Appendix

"Rules for the Study of Natural Philosophy" as presented in the third edition (1726) *Principia*

In the chapters that follow, I address the commentaries that accompany each of the four Rules. However, to maintain my line of discussion, I treat the various commentaries at different places in my presentation and do not examine the remarks that accompany Rule 3 in the order that they appear in the text. So that readers have a point of reference, below is how the Rules and their commentaries are presented in the third edition (1726) *Principia*.

I have added bracketed Latin terms to the two instances where my translation departs from what Cohen and Whitman present in Newton (1999). Namely, where I have "generally" and "general" in the commentary of Rule 3, they have "universally" and "universal." All other bracketed terms and phrases are taken from their translation.

Rule 1: No more causes of natural things should be admitted than are both true and sufficient to explain their phenomena.

As the philosophers say: Nature does nothing in vain, and more causes are in vain when fewer suffice. For nature is simple and does not indulge in the luxury of superfluous causes.

Rule 2: Therefore, the causes assigned to natural effects of the same kind must be, so far as possible, the same.

Examples are the cause of respiration in man and beast, or of the falling of stones in Europe and America, or of the light of a kitchen fire and the sun, or of the reflection of light on our earth and the planets.

Rule 3: Those qualities of bodies that cannot be intended and remitted [i.e., qualities that cannot be increased and diminished] and that belong to all bodies on which experiments can be made should be taken as qualities of all bodies universally.

For the qualities of bodies can be known only through experiments; and therefore qualities that square with experiments generally [*generales*] are to be regarded as general [*generaliter*] qualities; and qualities that cannot be diminished cannot be taken away from bodies. Certainly idle fancies ought not to be fabricated recklessly against the evidence of experiment, nor should we depart from the analogy of nature, since nature is always simple and ever consonant with itself. The extension of bodies is known to us only through our senses, and yet there are bodies beyond the range of these senses; but because extension is found in all sensible bodies, it is ascribed to all bodies universally. We know by experience that some bodies are hard. Moreover, because the hardness of the whole arises from the hardness of its parts, we justly infer from this not only the hardness of the undivided particles of bodies that are accessible to our senses, but also of all other bodies. That all bodies are impenetrable we gather not by reason but by our senses. We find those bodies that we handle to be impenetrable, and hence we conclude that impenetrability is a property of all bodies universally. That all bodies are movable and persevere in motion or in rest by means of certain forces (which we call forces of inertia) we infer from finding these properties in the bodies that we have seen. The extension, hardness, impenetrability, mobility, and force of inertia of the whole arise from the extension, hardness, impenetrability, mobility, and force of inertia of each of the parts; and thus we conclude that every one of the least parts of all bodies is extended, hard, impenetrable, movable, and endowed with a force of inertia. And this is the foundation of all natural philosophy. Further, from phenomena we know that the divided, contiguous parts of bodies can be separated from one another, and from mathematics it is certain that the undivided parts can be distinguished into smaller parts by our reason. But it is uncertain whether those parts which have been distinguished in this way and not yet divided can actually be divided and separated from one another by the forces of nature. But if it were established by even a single experiment that in the breaking of a hard and solid body, any undivided particle underwent division, we should conclude by the force of this third rule not only that divided parts are separable but also that undivided parts can be divided indefinitely.

Finally, if it is universally established by experiments and astronomical observations that all bodies on or near the earth gravitate [*lit.* are heavy] toward the earth, and do so in proportion to the quantity of matter in each body, and that the moon gravitates [is heavy] toward the earth in proportion to the quantity of its matter, and that our sea in turn gravitates [is heavy] toward the moon, and that all the planets gravitate [are heavy] toward one another, and that there is a similar gravity [heaviness] of comets toward the sun, it will have to be concluded by this third rule that all bodies gravitate toward one another. Indeed, the argument from phenomena will be even stronger for universal gravity than for the impenetrability of bodies, for which, of course, we have not a single experiment, and not even an observation, in the case of heavenly bodies. Yet I am by no means affirming that gravity is essential to bodies. By inherent force I mean only the force of inertia. This is immutable. Gravity is diminished as bodies recede from the earth.

Rule 4: In experimental philosophy, propositions gathered from phenomena by induction should be considered either exactly or very nearly true notwithstanding any contrary hypotheses, until yet other phenomena make such propositions either more exact or liable to exceptions.

This rule should be followed so that arguments based on induction may not be nullified by hypotheses.

(Newton 1999, 794–796)

2 The Rules in the Argument for Universal Gravity

2.1 Gravity as an Inverse-Square Force by Rules 1, 2, and 4

In the opening propositions of Book 3, Newton presents a series of claims concerning the motion of orbiting bodies around the bodies at the center of their orbits. He demonstrates each of these claims using propositions established in Book 1 and relevant astronomical observations, some of which are reported in the "Phenomena" section at the opening of Book 3. Initially, Newton focuses on the orbital motion of a particular set of celestial bodies around their respective centers: the moons of Jupiter around Jupiter (III.1), the moons of Saturn around Saturn (III.1), the planets around the sun (III.2), and the earth's moon around the earth (III.3). He shows in these first three propositions that the observed motions of each of these bodies can be explained by an inverse-square force, namely, by a centripetal (center-directed) force that varies inversely as the square of the distance between the orbiting body and its central body. In III.4 and III.5 and their corollaries, Newton establishes that the motions of these heavenly bodies can also be explained in reference to a force of gravity, that is, in reference to their tendency to fall toward their respective centers. In the Scholium to III.5, he additionally claims that gravity is both identical to and the cause of the force "by which celestial bodies are kept in their orbits" (Newton 1999, 806). In other words, according to Newton, the motions of celestial bodies can be explained by an inverse-square centripetal force precisely because they have a force of gravity – a tendency to fall toward their central body – that is governed by an inverse-square law.

To reach this general result, Newton first establishes in the proof to III.4 that the gravity of terrestrial bodies is an inverse-square force. He demonstrates this identity by using the results of III.3 to connect the fall of terrestrial bodies – the heavy bodies, which are on or near

DOI: 10.4324/9781003184256-2

the surface of the earth – with the motion of the moon. According to III.3, the moon's observed orbit around the earth can be explained by an inverse-square force. According to the corollary of III.3, the same inverse-square force would characterize the moon's motion if it were to fall to the earth (*ibid*, 802–803). In the proof to III.4, Newton focuses on just such a scenario – he imagines the moon "to be deprived of all its motion and to be let fall so that it will descend to the earth with all that force urging it by which (by prop. 3, corol.) it is [normally] kept in its orbit" (*ibid*, 804) – and calculates the rates of fall that would characterize the moon's motion in this circumstance.[26] He first determines that, as the moon begins its descent, "in the space of one minute, it will by falling describe $15^{1}/_{12}$ Paris feet" (*ibid*). He then determines the moon's rate of fall at the surface of the earth. Factoring in that the moon is falling according to a force that varies inversely as the square of its distance from the earth, he establishes that the force of the falling moon "at the surface of the earth is 60×60 times greater than at the moon" and calculates that, at the surface of the earth, it will take just one second for the moon to fall "$15^{1}/_{12}$ feet, or more exactly [to fall] 15 feet, 1 inch, and $1^{4}/_{9}$ lines" (*ibid*; bracketed phrase added).

Drawing on the results of pendulum experiments that had been conducted by Christiaan Huygens, Newton reports that terrestrial bodies fall at a nearly identical rate. Namely, in the span of one second, heavy bodies "in our regions" fall a distance of "15 Paris feet, 1 inch, and $1^{7}/_{9}$ lines" (*ibid*). And for Newton, this confirms that "heavy bodies do actually descend to the earth" with the same force as the falling moon, that is, that they descend with a force that varies inversely as the square of their distance from the earth's surface (*ibid*).

Newton's next step is to demonstrate that this inverse-square force of terrestrial bodies just is their force of gravity. To do so, he considers what would be observed if the two forces were not identical, and drawing again on Huygens's experimental results, he explains that:

> if gravity were different from this [inverse-square] force, then bodies making for the earth by both forces acting together would

---

26 Newton's calculations of the moon's rates of fall depend on the estimated values that he presents for the circumference of the earth, the distance between the earth and the moon, and the revolution of the moon with respect to the fixed stars (Newton 1999, 804). For discussion of Newton's treatment of the moon's rates of fall, and the possible fudging of his results, see Section 8.6, and especially Note 17, of Cohen (1999).

descend twice as fast, and in the space of one second would by falling describe $30\frac{1}{6}$ Paris feet, entirely contrary to experience.

(*ibid*, 804; bracketed term added)

The calculated result of $30\frac{1}{6}$ Paris feet is "entirely contrary to experience," because, as had already been noted, Huygens's pendulum experiments revealed that heavy bodies describe only "15 Paris feet, 1 inch, $1\frac{7}{9}$ lines" during one second of fall (*ibid*). Consequently, for terrestrial bodies, there is experimental justification for the claim that their force of gravity just is an inverse-square force. Each force can separately be used to explain the descent of these bodies, but the experimental observations of the space that they describe during their fall indicate that there is a single force, not two separate forces, acting on them as they descend to the earth.

In III.4 and III.5, Newton makes the same identification for celestial bodies; he establishes that the inverse-square force that explains their observed motions just is the force of gravity by which they fall toward their central bodies. However, the strategy that he uses in these cases is different than the one he uses for heavy bodies, and necessarily so. Celestial bodies are bodies on which experiments cannot be made, which means that no experiments on these bodies could be conducted to determine the space that they would describe if they fell toward the centers of their orbits. Accordingly, to establish that the inverse-square force that explains their observed motion is identical to their force of gravity, Newton extends to celestial bodies the identity that he had experimentally established for terrestrial bodies, and he does so by using the first, second, and fourth Rules for the Study of Natural Philosophy.

Rules 1 and 2 are explicitly referenced in the proof to III.4, where Newton identifies the inverse-square force of the earth's moon with the force of gravity. After establishing that the gravity of terrestrial bodies obeys an inverse-square law, he writes:

that force by which the moon is kept in its orbit, in descending from the moon's orbit to the surface of the earth, comes out equal to the force of gravity here on earth, and so (by rules 1 and 2) is that very force which we generally call gravity.

(*ibid*)

Newton moves quickly through the reasoning here, but his explicit reference to Rules 1 and 2 help clarify his justification for ascribing to the moon the same tendency to fall toward the earth – that is, the same force of gravity – that is found among terrestrial bodies. Generally speaking, Rule 1 directs us not to multiply the causes of natural phenomena unnecessarily. It tells us, more specifically, that "No more causes of natural things should be admitted than are **both true and sufficient to explain their phenomena**" (*ibid*, 794; boldface added). Courtesy of Huygens, Newton had experimental reasons for accepting that there is a single cause for the fall of terrestrial bodies and for accepting that this single cause was sufficient to explain their observed descent to the earth. No such direct experimental support could be marshalled to confirm that there is a single cause of the moon's observed motion. But with his use of Rule 1, Newton is claiming that no such direct experimental support is required. The evidence available indicates that a single cause – and specifically, a single force – suffices to explain why heavy bodies are observed to fall according to an inverse-square law. Consequently, lest we violate the directive of Rule 1, we should accept that a single cause will suffice to explain why the moon is observed to orbit the earth according to an inverse-square force.

The reason that we should ascribe the *same* single cause to the observed motion of the moon as has been ascribed to the observed motion of terrestrial bodies rests on Rule 2. This rule tells us that "the causes assigned to natural effects of the same kind must be, so far as possible, the same" (*ibid*, 795). So, just as we would not claim that what causes the fall of a stone in Europe is different than what causes the fall of a stone in America (at least according to Newton; *ibid*), we should not claim that the reason the moon moves according to an inverse-square force is different than the reason that terrestrial bodies fall according to an inverse-square force. Experiment has shown us that heavy bodies fall in this way because they are acted on by a single force – a force of gravity that varies inversely as the square of the distance between them and the earth's surface. Thus, as per Rule 2, we should say the same of the moon: It is acted on by a single force of gravity that varies inversely as the square of the distance between it and the earth's surface.[27]

---

27 Newton also references Rules 1 and 2 in the alternative proof for the identification of these forces that he presents in the scholium that accompanies III.4. In that proof, he considers what would occur if several moons revolved around the earth and shows that the descent of a moon that nearly touches the "tops of the highest mountains" should be explained in the same way as the descent of the terrestrial (heavy) bodies that are situated on the tops of those mountains. Namely, both should be taken to be caused by a single force of gravity that varies inversely as the square of their distance from the earth's surface. Applying the same single cause to all the several moons, Newton writes:

> since both forces – namely, those of heavy bodies and those of the moons – are directed toward the center of the earth and are similar to each other and equal

Newton applies similar reasoning about causes and effects in the Scholium to III.5, where he tells us that we should assign to the orbital motions of all celestial bodies the same cause that's been assigned to the orbital motion of the moon. Having shown in III.5 that the observed motions of the planets, and of the moons of both Jupiter and Saturn, can be explained in reference to a tendency to fall toward their respective centers, Newton states in the scholium:

> Hitherto we have called "centripetal" that force by which celestial bodies are kept in their orbits. It is now established that this force is gravity, and therefore we shall call it gravity from now on. For the cause of the centripetal force by which the moon is kept in its orbit ought to be extended to all the planets, by rules 1, 2, and 4.
>
> *(ibid,* 806)

The use of Rules 1 and 2 here is akin to what we saw in the proof of III.4. Having already established that a single cause is sufficient to explain why the observed orbital motion of the moon obeys an inverse-square law, following Rule 1, we should admit that a single cause also suffices to explain why other celestial bodies are observed to orbit their centers according to an inverse-square law. Following Rule 2, we should additionally accept that what causes the observed orbital motion of celestial bodies is the same as what causes the moon's observed orbit around the earth, because all these motions are "natural effects of the same kind," namely, they are all observed motions that can be explained by an inverse-square force.[28] Or, as Newton says in

---

[that is, since they are both inverse-square centripetal forces], they will (by rules 1 and 2) have the same cause. And therefore that force by which the moon is kept in its orbit is the very one that we generally call gravity.

(Newton 1999, 805; bracketed phrase added)

28   Newton makes explicit reference to this similarity in the proof to III.5, where he uses Rule 2 to establish that the inverse-square force that governs the orbital motions of the planets and their satellites is caused by their gravity, that is, by their tendency to fall toward their central bodies. He writes:

> For the revolutions of the circumjovial planets about Jupiter, of the circumsaturnian planets about Saturn, and of Mercury and Venus and the other circumsolar planets about the sun **are phenomena of the same kind** as the revolution of the moon about the earth, and **therefore (by rule 2) depend on causes of the same kind**, especially since it has been proved that the forces on which those revolutions depend are directed toward the centers of Jupiter, Saturn, and the sun, and decrease according to the same ratio and law (in receding from Jupiter, Saturn, and the sun) as the force of gravity (in receding from the earth).
>
> *(ibid,* 806; boldface added)

the Scholium to III.5, since it has already been shown that gravity is the "cause of the centripetal force by which the moon is kept in its orbit," by Rule 2 we ought to accept that gravity also causes the centripetal forces by which other "celestial bodies are kept in their orbits."

Additionally, we can make this generalizing move without being troubled by alternative explanations of these motions that are not as well supported by empirical evidence as the results that were established prior to the Scholium to III.5. This, at least, appears to be the general point that Newton is urging with his explicit reference to Rule 4 in the scholium. For, according to Rule 4:

> In experimental philosophy, propositions gathered from phenomena by induction should be considered either exactly or very nearly true **notwithstanding any contrary hypotheses**, until yet other phenomena make such propositions either more exact or liable to exceptions. (*ibid*, 796; boldface added)[29]

2.2 Weight as Directly Proportional to Mass by Rule 3

Once we reach the end of the Scholium to III.5, Newton has shown that, like terrestrial bodies, celestial bodies have a force of gravity. He has also shown that the gravity of both terrestrial and celestial bodies obeys an inverse-square law. He has thus established that the measure of gravity varies according to distance, namely, he has established that a body's gravity varies inversely as the square of its distance from the body to which it is attracted (or toward which it is falling). A couple of propositions later, Newton shows that a body's gravity also varies according to its mass. In the terms of III.7, he shows that "Gravity exists in all bodies universally and is proportional to the quantity of matter in each" (*ibid*, 810). To establish this relationship between gravity and mass, Newton first examines the relationship between the

---

29 That Rule 4 indicates that explanations lacking appropriate empirical support should be given no consideration when doing experimental philosophy is clear from the General Scholium. As noted in Section 1.1, there Newton asserts, "For whatever is not deduced from the phenomena must be called a hypothesis; and hypotheses, whether metaphysical or physical, or based on occult qualities, or mechanical, have no place in experimental philosophy" (*ibid*, 943). As for why in the third edition Newton adds the reference to Rule 4 in the Scholium to III.5, see Biener and Smeenk (2012). They provide a trenchant account of how the addition of Rule 4 provided Newton a way of replying (albeit not unproblematically) to questions raised by Roger Cotes about the legitimacy of extending gravity to bodies on which experiments cannot be performed.

weights and masses of particular sets of gravitating bodies, and it is in this context, and in his explanation of Corollary 2 of III.6 in particular, that he makes the one and only reference to Rule 3 in the argument for universal gravity.

The claims that Newton forwards in III.6 and its first and second corollaries are justified, in part, by results that he had gathered from a series of pendulum experiments, which he describes in the proof to III.6. In conducting these experiments, Newton compared the oscillations of pendulums that had bobs composed of different kinds of materials, and his general goal was to confirm that "the falling of all heavy bodies toward the earth … takes place in equal times" (*ibid*, 806). As he reports:

> I have tested [the equality of the times of fall] with gold, silver, lead, glass, sand, common salt, wood, water, and wheat. I got two wooden boxes, round and equal. I filled one of them with wood, and I suspended the same weight of gold (as exactly as I could) in the center of oscillation of the other. The boxes, hanging by equal eleven-foot cords, made pendulums exactly like each other with respect to their weight, shape, and air resistance. Then, when placed close to each other [and set into vibration], they kept swinging back and forth together with equal oscillations for a very long time. (*ibid*, 807; first bracketed phrase added)

According to Newton's description, the two pendulums that he initially assembled differed only with respect to the material composition of their bobs. The other salient features were the same. The bobs were equally distant from their pivots, they had identical weights and shapes, and they met with the same degree of air resistance during their oscillations. What the experiment revealed is that when the variously composed bobs of these pendulums were allowed to fall from the same height, they oscillated in equal times.

In Book 2, Newton had established that when simple pendulums in a vacuum oscillate in equal times, the weights of the bobs stand in direct proportion to their masses (*ibid*, 700; II.24, Corollary 1). He also established that the weight of the bob of any simple pendulum that moves in a nonresisting medium is directly proportional to the bob's mass (*ibid*, 701; II.24, Corollary 6). In the proof to III.6, Newton applies these results to the initial findings of his pendulum experiments and establishes that the masses of the gold and wood bobs are directly proportional to their respective weights, that is, that the mass of the gold is to the mass of the wood as the weight of the gold is to the weight of the wood.[30] He then reports that he reached the same result when he compared the motions of pendulums that had bobs made of other materials. In all of these cases, the weights of the bobs were directly proportional to their quantities of matter (*ibid*, 807).[31]

After reporting this result, Newton uses the same kind of reasoning that he did in the proof to III.4: He reasons from a counterfactual situation to extend to celestial bodies the experimentally supported conclusion that he had established for terrestrial bodies. In III.4, he considered a situation in which the moon was deprived of its orbital motion and allowed to fall to the earth. In the proof to III.6, he instead considers a situation in which "our terrestrial bodies [are] raised as far as the orbit of the moon and, together with the moon, deprived of all motion, [and] released so as to fall to the earth simultaneously" (*ibid*, 807; bracketed terms added). He has proposed a scenario in which various heavy bodies start their fall to the earth from the same height, and having already shown that all these bodies tend to fall toward the earth according to an inverse-square force, it is a scenario in which all the bodies will fall to earth in the same time. In turn, it is a scenario to which the results of the pendulum experiments can be applied, which is why Newton concludes that when terrestrial bodies and the moon fall to the earth from the same height, there is a direct proportionality between the weights and masses of the terrestrial bodies and the weight and mass of the moon (*ibid*).

As the proof to III.6 continues, Newton uses the same general strategy in his treatment of the fall of other celestial bodies toward their centers (e.g., of the moons of Jupiter toward Jupiter and of the circumsolar planets toward the sun) and in his treatment of "the weights [or gravities] of the individual parts of each planet toward any other planet" (*ibid*, 808). Through this iterative process, which began from

---

30 In Newton's words,

> Accordingly, the amount of matter in the gold (by book 2, prop. 24, corols. 1 and 6) was to the amount of matter in the wood as the action of the motive force upon all the gold to the action of the motive force upon all the [added] wood – that is, as the weight of one to the weight of the other.
>
> (*ibid*, 807)

31 For discussion of the connection between the experiments that are presented in the *Principia* and the mathematical principles that Newton establishes to explain natural phenomena, see Chapter 9 of Bertoloni Meli (2006).

the results of his pendulum experiments, Newton marshals the support he needs to support the general claim of III.6, namely, that

> All bodies gravitate toward each of the planets, and at any given distance from the center of any one planet the weight [*pondera*] of any body whatever toward that planet is proportional to the quantity of matter which the body contains.
>
> (*ibid*, 806; bracketed term added)

The pendulum experiments also help justify Corollary 1 of III.6, which states that "the weights [*pondera*] of bodies do not depend on their forms and textures" (*ibid*, 809; bracketed term added).[32] To show that there is no such dependence, Newton presents a one-sentence *reductio* argument in which he considers what would occur if weight did depend on the forms of bodies. In such a circumstance, the weights of bodies that have equal mass would vary according to their "variety of forms." But this, Newton notes, is "entirely contrary to experience." Specifically, it is entirely contrary to what Newton had observed in his pendulum experiments. He had created bobs using equally massive bodies of different forms – bodies made of various materials including gold, wood, silver, and wheat – and there was no noticeable relationship between the weights of these bobs and the various materials of which they were composed. Their weights varied only according to their masses.

Newton focuses on two other results from his pendulum experiments to establish Corollary 2 of III.6:

> All bodies universally that are on or near the earth are heavy [or gravitate; *gravia*] toward the earth, and the weights [*pondera*] of all bodies that are equally distant from the center of the earth are as the quantities of matter in them.
>
> (*ibid*, 809; bracketed Latin terms added)[33]

---

32 While not put in precisely these terms, the claim that weight varies by a body's form and texture is defended by Descartes in Part IV of the *Principles*. See Section 1.1.

33 "Corpora universa, quæ circa terram sunt, gravia sunt in terram; & pondera omnium, quæ æqualiter a centro terræ distant, sunt ut quantitates materiæ in iisdem" (Newton 1871, 402). In the first edition (1687), the statement of Corollary 2 begins with "Therefore"; otherwise, its presentation is identical in all three editions of the *Principia*. Given Newton's phrasing, there is a slight ambiguity surrounding the term "omnium." Looking at the corollary in isolation, the term could be read as qualifying bodies in general or as qualifying the "bodies on or near the earth" that are referenced in the first clause. That it makes most sense to read the "omnium" as

In conducting the pendulum experiments, Newton used several kinds of bodies – the bobs were made of "gold, silver, lead, glass, sand, common salt, wood, water, and wheat" – and each kind of body had a heaviness, or weight, relative to the earth. In the first clause of Corollary 2 of III.6, Newton is extending this basic feature to all terrestrial bodies. He is claiming that, like the various types of bodies used in the experiment, all bodies on or near the earth have a heaviness, and thus a measurable weight, relative to the earth.[34]

The second clause of the corollary concerns the weight of a particular set of terrestrial bodies, namely, those that are equally distant from the center of the earth. Here, Newton is alluding to a different feature of the bobs in the pendulum experiments, specifically, that all of them had been hung at an equal distance – "by equal eleven-foot cords" – from their pivots. At these equal distances he found that, for all the types of bodies that he used, the weights of the bobs varied directly with their masses, or quantities of matter. In the second clause of the corollary, Newton universalizes this proportionality to all terrestrial bodies that are equally distant from the center of the earth, and he is justified in doing so, he says, because "This is a quality of all bodies on which experiments can be performed and therefore by rule 3 is to be affirmed of all bodies universally" (*ibid*, 809).[35]

---

qualifying "bodies on or near the earth" is strongly suggested by the experimental results on which the corollary is based, as I discuss immediately below.

34 I read the "gravia" that Newton uses here in terms of heaviness and weight, rather than in terms of a general tendency to fall to the earth, for two reasons. This reading is consistent with what is signaled by the experiments described in the proof to III.6. Additionally, in the proof to III.4, and in reference to pendulum experiments conducted by Huygens, Newton had already forwarded the claim that terrestrial bodies have a tendency to fall to earth according to an inverse-square force. There would be no need for Newton to repeat that claim here, let alone repeat it as part of a corollary of III.6.

35 "Hæc est qualitas omnium in quibus experimenta instituere licet, & propterea per reg. III de universis affirmanda est" (Newton 1871, 402). This sentence is added to the second edition of the *Principia* – the same edition in which Rule 3 first appears – and it is retained in the third edition without modification. In all three editions, the remainder of the commentary that accompanies Corollary 2 of III.6 is nearly identical. The most significant change is that, in the second and third editions, Newton describes the possible transmutation of bodies as "the opinion of Aristotle, Descartes and others," whereas in the first edition, he had presented this possible transmutation at the opening of Book 3 as Hypothesis III. There is debate about what prompted Newton to make this change. According to Cohen (1966), Newton's replacement of Hypothesis III with Rule 3 indicates Newton's "personal disbelief" in the ontological claim of Hypothesis III. In contrast, McGuire (1968) posits that Newton removed Hypothesis III from the second edition *Principia*,

Recall what Rule 3 says:

> Those qualities of bodies that cannot be intended and remitted
> and that belong to all bodies on which experiments can be made
> should be taken as qualities of all bodies universally.
>
> *(ibid, 795)*

The "all bodies universally" referenced in the Rule includes all the
natural bodies that we have yet to observe and also all the natural
bodies that are "beyond the range of [our] senses," as Newton explains
in the commentary accompanying the Rule (*ibid*; bracketed term
added). Consequently, since the second clause of Corollary 2 of III.6
has been established "by rule 3," the same meaning of "all bodies universally" applies there. That is, the second clause communicates that
the proportionality between weight and mass is a quality of all sensible
and insensible terrestrial bodies that are equidistant from the center
of the earth.[36]

As for why Rule 3 can be applied to this proportionality, Newton
explicitly reports that it is because it meets one of the two conditions communicated by the Rule. The proportionality is, he says, "a
quality of all bodies on which experiments can be performed" (*ibid*,
809). You'll notice that this explanation rests on a generalization that
Newton does not address or defend in his presentation of Corollary
2 of III.6. With his pendulum experiments he had gathered evidence
that the proportionality between weight and mass was a quality of the
bodies on which he *did* conduct experiments; it was a quality of the
bobs that were hung from their pivots "by equal eleven-foot cords."
What he reports immediately above is that this proportionality is a

---

not because he rejected the ontological position expressed in that Hypothesis, but
because "he probably came to the conclusion that the whole problem-area [of the
mutual transformation of bodies] was too complex for an official solution which
could be set forth briefly in the *Principia*" (McGuire 1968, 235; bracketed phrase
added). On this question, see also Biener (forthcoming, Section 1.3), who suggests
that Newton replaced Hypothesis III with Rule 3 to address Huygens's worry that
the tendency of bodies to fall toward other bodies was an effect of a not-yet identified mechanical cause, and in turn, a tendency that could not be generalized to all
bodies until the unknown cause had been discovered.

36 I use "bodies" here in a broad sense that includes the parts of bodies. In Section 4.1,
I take a more careful look at Newton's justification for generalizations that proceed
from what is sensible to what is insensible. In Section 3.3, I discuss Newton's example of an experiment that allows us to generalize the quality of divisibility from one
sensible part of a body to all the sensible parts of all bodies.

quality of *all* experimental bodies that are equally distant from the center of the earth – that it is "a quality of all bodies on which experiments **can be performed**" (*ibid*, 809; boldface added). Newton made the same sort of inductive move earlier in Book 3, and also did so without offering an explanation or defense. Namely, in the proof to III.4, Newton had generalized the results of Huygens's pendulum experiments to justify the claim that all terrestrial bodies fall toward the earth according to an inverse-square force. Similarly, Newton used his own pendulum experiments as the basis for the general claim, made in the first clause of Corollary 2 of III.6, that "All bodies universally that are on or near the earth are heavy [or gravitate] toward the earth." In each of these cases, Newton proceeds from claims about those bodies on which experiments *have been made* to claims about all bodies on which experiments *can be made*, that is, from claims justified by *some* experimental bodies to claims that range over *all* experimental bodies. Or, in the terms Newton uses in the General Scholium, he forwards propositions that have been "deduced from the phenomena and ... made general by induction" (*ibid*, 943). That he offers no explanation or defense when he makes these sorts of inductive generalizations suggests that he finds (and thinks his readers would find) such generalizations unproblematic and uncontroversial. Indeed, what his argument for universal gravity bears out is that, for Newton, only generalizations that extend from a particular set of experimental data to all bodies *universally* require special justification, and the application of Rule 3, in particular.[37]

Depending on how one reads Rule 3, in presenting Corollary 2 of III.6, Newton is also taking for granted that the proportionality between weight and mass is a quality of bodies that "cannot be intended and remitted." He does not explicitly say this in the corollary, or elsewhere in the *Principia*. But if we read Rule 3 as expressing the two necessary and sufficient conditions that must be met for a quality of bodies to be universalized, then the Rule can be applied to the proportionality between weight and mass if and only if this proportionality is a quality that can be found among all bodies on which experiments

---

37 Earlier in the text, Newton does offer a general explanation for why the experimental philosopher is justified in making generalizations that proceed from bodies on which experiments *have been made* to bodies on which experiments *can be made*. Namely, in the commentary accompanying Rule 3, he suggests that it is permissible to make these sorts of some-to-all generalizations when reasoning about experimental evidence, because "nature is always simple and ever consonant with itself" (*ibid*, 795). I discuss this justification in Section 3.3.

can be performed *and* is also a quality that "cannot be intended and remitted." In other words, on the One-Set Reading of Rule 3, just in virtue of the fact that Newton applies the Rule to the proportionality between the weights and masses of bodies that are equidistant from the center of the earth, he must accept that the proportionality has both of these features.[38]

I noted in Section 1.2 that I favor a different reading of Rule 3. I take it to be a statement about two sets of qualities, not one. Relatedly, I do not think that in Corollary 2 of III.6 Newton is assuming that the proportionality between weight and mass is a quality that "cannot be intended and remitted." But before I move on in Chapter 3 to defend the plausibility of my Two-Set Reading, it is worth taking stock of what follows from applying the more standard One-Set Reading of Rule 3 to Corollary 2 of III.6.

2.3 The One-Set Reading and Its Consequences

Accepting that Rule 3 communicates the two necessary and sufficient conditions that the qualities of bodies must meet for them to be taken as qualities of all bodies universally is consistent with the text of Corollary 2 of III.6. It also provides us a reading that underscores the importance of experimental evidence in Newton's reasoning. As we just saw, Newton explicitly states that the proportionality between weight and mass meets one of the two conditions communicated by Rule 3. It is, as he says, "a quality of all bodies on which experiments can be performed" (*ibid*, 809). Moreover, with his pendulum experiments, Newton had established that the proportionality between the weights and the masses of the bobs was stable. That is, he discovered that the measures of these qualities stood in direct proportion no matter which specific materials he used to construct the bobs and no matter which specific values of weight and mass he had measured. Consequently, on the assumption that the set of qualities that "cannot be intended and remitted" includes such invariant proportions, Newton had experimental evidence that the proportionality between weight and mass is a quality that meets the other condition expressed by Rule 3.

Now, to make this particular assumption is to read Newton as departing from the standard medieval usage of *intendi & remitti*.

---

38 As mentioned in Section 1.2, Cohen (1999) explicitly endorses the One-Set Reading of Rule 3 (cf. Cohen 1999, 199). The One-Set Reading is also adopted by Biener (forthcoming), Okruhlik (1989), Janiak (2008), and Ducheyne (2012) in their respective treatments of Newton's use of Rule 3 to establish Corollary 2 of III.6.

As noted in Section 1.2, according to the medieval theory of the latitude of forms, a quality can be "intended or remitted" if and only if it is a quantifiable quality that can increase or decrease.[39] And according to standard medieval usage, the question of whether a quality could be intended and remitted was a question of whether the specifiable measures of the quality varied in degree. It was on this basis that qualities such as motion, displacement, cold, and heat were considered qualities that could be intended and remitted (Cohen 1999, 200). To count the invariant *proportionality* between weight and mass as a quality that cannot be intended and remitted, as Newton is allegedly doing in Corollary 2 of III.6, means that the standards for identifying such qualities have changed. It is not a question of whether a particular quality has a specifiable measure that can increase or decrease. The question instead is whether the mathematical relationship between two quantifiable qualities remains stable. It is in this respect that the proportionality between weight and mass can be counted as a quality that cannot be intended and remitted: It is a quantifiable, mathematical relation that remains invariant, even as the specific measures of a body's weight and mass are increased or decreased.[40]

For those who adopt the One-Set Reading of Rule 3, there is an additional advantage to interpreting Newton's use of *intendi & remitti* along these non-medieval lines. It provides a fruitful way of reconciling the variation of gravity's measure with Newton's claim that Rule 3 can be used to universalize gravity. The reason that questions surround this particular claim is because Newton establishes in Book 3 that the force of gravity is governed by an inverse-square law, that is, he establishes that the specific measure of a body's gravity will vary according to the body's distance from the body to which it is attracted. The farther that a terrestrial body is from the surface of the earth, for instance, the lower its weight relative to the earth. Or, as Newton puts the general point, "Gravity is diminished as bodies recede from the earth" (*ibid*, 796). Consequently, if we adopt the standard medieval usage of *intendi & remitti*, gravity is a quality of bodies that *can* be

---

39 For evidence that Newton's use of *intendi & remitti* in the statement of Rule 3 signaled his appropriation of this medieval theory, see the works cited in Note 24.

40 My characterization of how Newton's use of *intendi & remitti* might depart from the standard medieval usage is consistent with the interpretations offered by Okruhlik (1989) and Janiak (2008), and with the position explicitly defended in Ducheyne (2012, Section 3.2). Belkind (2017) provides an alternative reading of the non-medieval sense in which Newton uses *intendi & remitti* that is based on the atomist commitments expressed in the commentary to Rule 3 (cf. Note 63).

intended and remitted, and in turn, it is a quality that fails to meet one of the two conditions expressed by Rule 3.

Reading Newton's use of *intendi & remitti* along the non-medieval lines sketched above yields a different result. What's important here is that even as the specific measures of a body's gravity vary (as its distance from the body to which it is attracted varies), there are mathematical relations that remain stable. Namely, under all circumstances, the measure of a body's gravity will vary directly with the body's mass, and, under all circumstances, its measure will vary inversely as the square of the distance between the body and the body to which it is attracted. These mathematical relationships that gravity bears to mass and distance are invariant, which means that the *general* measure of gravity is characterized by proportions that do not vary in degree. Consequently, on the non-medieval reading of *intendi & remitti*, gravity can be counted as a quality of bodies that cannot be intended and remitted.[41]

We thus have from this reading a way to make sense of Newton's claim that gravity could be universalized on the basis of Rule 3. In the commentary that accompanies the Rule, he states:

Finally, if it is universally established by experiments and astronomical observations [1] that all bodies on or near the earth

---

41 In her treatment of the sense in which gravity cannot be intended and remitted, Okruhlik (1989) notes that Newton uses "gravity" sometimes to refer to a force, that is, to "something with causal efficacy," and at other times to refer to weight, or *pondus*, that is, to the effect of the force of gravity (Okruhlik 1989, 111). On her account, in Corollary 2 of III.6, Newton is referring to gravity as a force (of mutual attraction) that is defined by a proportionality to mass, and he is referring to gravity as an effect when he says that "Gravity is diminished as bodies recede from the earth." My summary is consistent with Okruhlik's interpretation, but I opt to use the distinction between *the general measure of gravity* and *the specific measures of a body's gravity* to clarify the sense in which the invariant proportions associated with gravity render it a quality that cannot be intended and remitted. Janiak (2008) takes a related but importantly different tack. As Okruhlik (1989) and Ducheyne (2012), Janiak accepts the One-Set Reading of Rule 3 and also emphasizes the invariant proportions that characterize gravity's general measure. However, he claims that Newton does not consider gravity to be quality of bodies. As Janiak has it, Newton's gravity is "a type of interaction rather than a quality," such that to say that gravity is universal is to say that it is a "universal type of interaction," that is, it is to say that "all bodies in the universe bear gravitational interactions (and indeed, with all other bodies)" (Janiak 2008, 96). Such an interpretation appears to be in tension with Newton's claim (discussed immediately below) that gravity could be universalized by Rule 3, since that Rule that clearly concerns qualities of bodies. Janiak addresses the tension in Janiak (2008, Chapter 4).

gravitate [*lit.* are heavy] toward the earth, and do so in proportion to the quantity of matter in each body, and [2] that the moon gravitates [is heavy] toward the earth in proportion to the quantity of its matter, and [3] that our sea in turn gravitates [is heavy] toward the moon, and [4] that all the planets gravitate [are heavy] toward one another, and [5] that there is a similar gravity [heaviness] of comets toward the sun, it will have to be concluded by this third rule that all bodies gravitate toward one another.

(*ibid*, 796; bracketed numbers added)

Newton is clearly emphasizing that, for Rule 3 to be applied to gravity, there must be experimental and observational evidence that gravity belongs to a host of different bodies. In turn, he is indicating that there must be sufficient evidence to merit considering gravity a quality that belongs "to all bodies on which experiments can be made."[42] He also notes that the gravity that is to be identified through experiment and observation is a force that is "in proportion to the quantity of matter in each body." This suggests that the evidence that's collected must show that the *general* measure of gravity is defined by an invariant proportion, and specifically, that its general measure always bears the same mathematical relation to mass. Newton can thus be understood as claiming that gravity can be universalized by Rule 3 only if experimental and observational evidence shows that gravity meets both of the conditions that the Rule identifies – if, that is, there is evidence that gravity is *both* a quality that belongs "to all bodies on which experiments can be made" *and* a quality that "cannot be intended and remitted." And on this reading, the variation in the *specific* measures of a body's gravity has no bearing on Newton's argument.[43]

---

42 Newton's remarks additionally indicate that the application of Rule 3 requires making the same sort of some-to-all generalization that was discussed in Section 2.2. That is, he's indicating that to apply Rule 3 to gravity, we have to generalize the gravity that's been found to belong to bodies on which experiments and observations *have been made* and consider it a quality of bodies on which experiments and observations *can be made.*

43 It is worth noting that Newton's summary of the steps that are required to apply Rule 3 to gravity does not map on to the literal way that the argument for universal gravity progresses in Book 3. For instance, Newton does not treat the motions of the tides or the motions of the comets until after he demonstrates in III.7 that "Gravity exists in all bodies universally and is proportional to the quantity of matter in each" (Newton 1999, 810). (The tides are treated in III.24 and III.36–37 and the comets in the section that ends Book 3.) Also, according to some of Newton's remarks, it seems that the theory of heavenly motions is meant to support his

This characterization of the sense in which gravity is a quality that cannot be intended and remitted brings with it a further advantage. As emphasized by Ducheyne (2012), it illuminates the specific way in which Newton takes gravity to be "of a different kind from the magnetic force" (Newton 1999, 810). This claim is presented as Corollary 5 of III.6, and in the commentary that follows, Newton explains:

> the magnetic force in one and the same body can be intended and remitted [i.e., increased and decreased] and is sometimes greater in proportion to the quantity of matter than the force of gravity; and this force, in receding from the magnet, decreases not as the square but almost as the cube of the distance, as far as I have been able to tell from certain rough observations.
>
> (*ibid*)

According to Newton's report, the measure of magnetic force appears to bear some relationship to distance; it decreases "almost as the cube of the distance," he says. But he has not identified any well-defined invariant proportions that characterize this measure. Specifically, he has found that the measure of magnetism does not always vary directly with mass and that it does not always vary inversely to the square of the distance between the attracting and attracted bodies. Consequently, adopting the reading above, Newton is claiming that the "force of gravity is of a different kind from the magnetic force" precisely because, having no specific invariant proportions that characterize its measure, magnetic force "can be intended and remitted," whereas gravity cannot.[44]

---

theory of comets, not vice versa. For instance, he says that the theory of comets "that observes the same laws as the planets, and that agrees exactly with exact astronomical observations cannot fail to be true" (*ibid*, 916). (I discuss the context and Latin construction of this statement in Section 3.2.) For a way of tracking the steps of the argument for universal gravity according to which Newton's treatments of the tides and comets are offered as support for the theory of universal gravity, see Smith (2016).

44 As additional evidence that Newton considered gravity to be a quality that cannot be intended and remitted, Ducheyne relies on manuscript material that predates the second edition *Principia* in which Newton says that the gravity of terrestrial bodies "is not intended and remitted" (Ducheyne 2012, 117). Compare this with McGuire (1968), who claims that the absence of any such remark about gravity in the second and third editions of the *Principia* indicates that Newton abandoned this position. McGuire supports his reading by noting that, elsewhere in the draft materials from the 1690s, Newton crossed out a passage in which he had written that gravity cannot be intended and remitted (McGuire 1968, 234).

In conjunction, then, with the non-medieval reading of Newton's use of *intendi & remitti*, the One-Set Reading of Rule 3 illuminates several important passages from Book 3. It provides a way of clarifying Newton's distinction between gravity and magnetism, Newton's application of Rule 3 in Corollary 2 of III.6, and Newton's claim that, when appropriate evidence is available, the Rule can be used to universalize gravity. However, when the One-Set Reading is applied to two other texts from Book 3, both of which are included in the commentary that accompanies the third Rule, questions arise about the cogency of Newton's stated position.

In the first case, Newton claims that, by Rule 3, we could accept that divisibility is a quality of all bodies universally "if it were established **by even a single experiment** that in the breaking of a hard and solid body, any undivided particle underwent division" (*ibid*, 796; boldface added). The puzzle here concerns the condition stated in Rule 3 that the universalizable qualities of bodies are ones that "belong to all bodies on which experiments can be made" (*ibid*, 795). In the accompanying commentary, Newton explains that among these "general qualities" are those that have been found to "square with experiments generally" (*ibid*). Surely, given the circumstances of Newton's example, divisibility cannot be considered a "general quality" on these grounds; the experimental philosopher has found divisibility to square only with the single experiment that she has performed. Consequently, if we adopt the One-Set Reading, and accept that Rule 3 is to be applied only to those qualities of bodies that meet both the conditions that the Rule identifies, the question that remains is why and in what sense a quality that is found to belong to a lone experimental body can be considered a quality that belongs to all experimental bodies.

The other puzzle that lingers for the One-Set Reading concerns the distinction that Newton explicitly draws between inertia and gravity in the third edition *Principia*. The basic claim he makes is that inertia can be considered an inherent force that is essential to bodies, whereas gravity cannot. In his terms,

> Yet I am by no means affirming that gravity is essential to bodies. By inherent force I mean only the force of inertia. This is immutable. Gravity [*Gravitas*] is diminished as bodies recede from the earth.
>
> (*ibid*, 796; bracketed term added)[45]

---

45 "Attamen gravitatem corporibus essentialem esse minime affirmo. Per vim insitam intelligo solam vim inertæ. Hæc immutabilis est. Gravitas recedendo a terra diminuitur" (Newton 1871, 389).

Based on the image of gravity that emerges from the One-Set Reading, there is a way to get some purchase on the distinction that Newton is drawing here. As we've just seen, on that reading gravity is a quality that cannot be intended and remitted insofar as its general measure is characterized by two invariant proportions. And as per the relation that is expressed by one of those proportions, the measure of a body's gravity will, under all circumstances, depend on the body's distance from the body to which it is attracted. This is not a feature that characterizes a body's inertia. The measure of a body's inertia does not depend on how the body is related to other bodies; it depends only on the body's mass.[46] Newton thus seems to have some legitimate grounds for claiming that inertia should be considered inherent and essential and gravity should not.

A question lingers nonetheless. If Newton is suggesting that inertia can be considered inherent and essential because it is immutable – and, in particular, because the specific measure of inertia for one and the same body will not vary – this implies that inertia can be considered inherent and essential because of the very invariance that presumably allows us to consider it a quality that cannot be intended and remitted. Gravity is not characterized by the same sort of invariance as inertia – gravity's invariance, or "immutability," is linked with the invariant *proportions* that characterize its general measure. However, on the One-Set Reading, this is the invariance that allows gravity to be considered a quality that cannot be intended and remitted. And so, Newton is left with the question: Why does the invariance that allows us to consider gravity a quality that cannot be intended and remitted not also one that allows us to consider gravity an inherent and essential quality of bodies? Or, put differently: Why does the sort of invariance that is associated with inertia, but not the sort that is associated with gravity, provide us grounds for considering a quality of bodies to be inherent in and essential to bodies?

I explore the broader significance of this question in Section 4.2. In Section 3.3 I consider how proponents of the One-Set Reading might

---

46 As Newton puts it in Definition 3,

> *Inherent force of matter is the power of resisting by which every body, so far as it is able, perseveres in its state either of resting or of moving uniformly straight forward.* This force is always proportional to the body and does not differ in any way from the inertia of the mass except in the manner in which it is conceived.
>
> (Newton 1999, 404)

I elaborate on the relevance of Definition 3 for the distinction that Newton draws between inertia and gravity in Section 4.2.

interpret Newton's example of universalizing divisibility. In these sections, I also argue that the Two-Set Reading of Rule 3 offers a way around the puzzles that emerge when the One-Set Reading is applied to these texts. First, though, in Section 3.1 I clarify what's at stake in the debate between the One-Set and Two-Set Readings of Rule 3 and begin my defense of the Two-Set Reading by showing that it is consistent with the same passages that can be clarified by the One-Set Reading.

3 The Two-Set Reading of Rule 3

3.1 Two Sets versus One

According to Rule 3, "Those qualities of bodies that cannot be intended and remitted [*intendi & remitti*] and that belong to all bodies on which experiments can be made should be taken as qualities of all bodies universally" (Newton 1999, 795). We saw in Chapter 2 that commentators have raised questions about the scope of the "cannot be intended and remitted" in the first clause and, specifically, about whether it includes qualities that are invariant proportions and qualities with measures that are characterized by invariant proportions. But remaining agnostic about what the *intendi & remitti* is meant to convey still leaves intact the general meaning of Rule 3. The Rule identifies two conditions that are to be met by the qualities of bodies that the experimental philosopher should regard as universal qualities.

What is not immediately clear is whether the Rule is stating that a universalizable quality must meet both conditions or that it need meet only one. In other words, there is an ambiguity about whether the two conditions identified in Rule 3 are necessary and sufficient or sufficient but not necessary. The ambiguity lingers, because it is not immediately clear whether the Rule refers to one set of qualities or two sets. If the Rule communicates that there is a single set of qualities of bodies that "should be taken as qualities of all bodies universally" – that is, that should be taken to belong to the general class of universalizable qualities – then the two conditions are necessary and sufficient. Alternatively, if the Rule communicates that there are two different but possibly overlapping sets of qualities of bodies that should be universalized, then the conditions are sufficient but not necessary. In somewhat plainer terms, Rule 3 lends itself to two possible interpretations.

DOI: 10.4324/9781003184256-3

The One-Set Reading of Rule 3: The Rule identifies a single set of qualities that ought to be universalized, and it communicates that the members of this single set are qualities of bodies both that "cannot be intended and remitted" and that "belong to all bodies on which experiments can be made."

The Two-Set Reading of Rule 3: The Rule identifies two sets of qualities that ought to be universalized, and it communicates that the members of one set are qualities of bodies that "cannot be intended and remitted" and that the members of the other set are qualities that "belong to all bodies on which experiments can be made."[47]

If we adopt the One-Set Reading and accept that Rule 3 expresses two necessary and sufficient conditions for membership in the general class of universalizable qualities, then, as we saw in Section 2.3, we can make sense of a variety of texts from Book 3. The Two-Set Reading shares this advantage, but it provides a different picture of the meaning of those texts and of the assumptions that support them.

Returning first to the single instance at which Newton applies Rule 3 in the argument for universal gravity, recall that, according to Corollary 2 of III.6:

All bodies universally that are on or near the earth are heavy [or gravitate; *gravia*] toward the earth, and the weights [*pondera*] of all bodies that are equally distant from the center of the earth are as the quantities of matter in them.

(Newton 1999, 809; bracketed Latin terms added)

Immediately after presenting this claim, Newton explains that the proportionality between weight and mass has been universalized to all bodies equidistant from the earth's center, because "This is a quality of all bodies on which experiments can be performed and therefore by rule 3 is to be affirmed of all bodies universally" (*ibid*). As noted

---

47 According to both of the readings I present here, Rule 3 is understood as offering a description of the evidentiary circumstances under which the experimental philosopher can justifiably infer that a quality belongs to all bodies universally. Consequently, on both readings, it is possible that natural bodies have universal qualities beyond those that the available evidence allows the experimental philosopher to identify. It is also possible that the qualities that have been universalized by means of the evidence that is specified in Rule 3 are not actually and in fact the universal qualities of bodies.

in Section 2.2, Newton's explanation for why Rule 3 can be applied to this particular case rests on an implicit generalization. From the pendulum experiments described in the proof of III.6, Newton identified a proportionality between the weights and masses of bobs hanging at equal distances from their pivots. What he says above, in the presentation of the proposition's second corollary, is that this proportionality "is a quality of all bodies on which experiments **can be** performed" (*ibid*, 809; boldface added), which means that he has generalized his experimental results. He has taken a quality discovered among *some* experimental bodies of a particular type and ascribed it to *all* experimental bodies of the same type. And it is this generalization that allows him to apply the third Rule and thereby extend the proportionality between weight and mass to all bodies *universally* that "are equally distant from the center of the earth."

Whether there is an additional assumption at play here depends on how many sets of qualities Rule 3 is identifying as universalizable. Again, according to the One-Set Reading, Rule 3 identifies a single set of universalizable qualities and communicates the two necessary and sufficient conditions that are met by the members of this set. If we adopt this reading, then, when Newton applies the Rule in Corollary 2 of III.6, he is taking for granted that the proportionality between the weights and masses of bodies that are equidistant from the earth's center is a quality that "cannot be intended and remitted." Nowhere in the *Principia* does Newton explicitly make this claim. But on the One-Set Reading, unless the proportionality meets both of the conditions expressed by Rule 3, it could not be universalized by use of the Rule. And assuming Newton does accept that the proportionality between weight and mass "cannot be intended and remitted," then, as we saw in Section 2.3, he has departed from medieval tradition and is using that category to pick out qualities that are invariant proportions.

The Two-Set Reading of Rule 3 yields a different result. According to this reading, Rule 3 identifies two sets of qualities and tells us that, to be universalized, a quality need belong only to one of those sets. In other words, the Rule indicates that a universalizable quality is a quality of bodies that evidence indicates is *either* one that "cannot be intended[and remitted" *or* one that "belong[s] to all bodies on which experiments can be made." Evidence could very well indicate a quality's membership in both sets. (I discuss such a circumstance in Section 4.2.) But evidence of membership in one set is all that is required for us to take a quality to belong to all bodies universally. On this reading, when Newton reports that Rule 3 can be applied to the proportionality between weight and mass in his presentation of

Corollary 2 of III.6, he is reporting that the proportionality is a member of one set of universalizable qualities. It is, as he says, "a quality of all bodies on which experiments can be performed" (*ibid*). And on the Two-Set Reading, this is all that he needs to say to justify his use of Rule 3, because on the Two-Set Reading, the proportionality's membership in a single set is all that Rule 3 requires. Consequently, the question of the precise meaning Newton has assigned to *intendi &* *remitti* is not one that needs to be addressed in this particular context.

The Two-Set Reading also allows us to bypass the apparent tension that is prompted by Newton's claim that Rule 3 could be used to universalize gravity. The tension emerges for the One-Set Reading, because by assuming that Rule 3 expresses the two necessary and sufficient conditions that are to be met by universalizable qualities, it appears that Newton holds two competing commitments about the sort of quality that gravity is. Insofar as he states that the Rule can be used to universalize gravity, he is claiming that gravity is a quality of bodies that *cannot* be intended and remitted. Yet, according to the medieval usage of *intendi & remitti*, he also affirms that gravity *can* be intended and remitted, insofar as he maintains that the specific measure of a body's gravity does increase and decrease, depending on the body's distance from the body to which it is attracted. Of course, this tension is not necessarily insoluble. We saw in Section 2.3 that proponents of the One-Set Reading have a way to reconcile Newton's apparently contradictory commitments.

But on the Two-Set Reading, the tension simply does not arise. On this reading, there is no need to assume that applying Rule 3 to gravity requires there be evidence that gravity is a quality that cannot be intended and remitted. So long as evidence shows that gravity is a quality that belongs "to all bodies on which experiments can be made," gravity could be universalized on the basis of the Rule. And, on the Two-Set Reading, this is precisely what Newton is emphasizing in his explanation of the evidentiary circumstances under which Rule 3 could be applied to gravity. Recall what he says:

Finally, if it is universally established by experiments and astronomical observations [1] that all bodies on or near the earth gravitate [*lit*. are heavy] toward the earth, and do so in proportion to the quantity of matter in each body, and [2] that the moon gravitates [is heavy] toward the earth in proportion to the quantity of its matter, and [3] that our sea in turn gravitates [is heavy] toward the moon, and [4] that all the planets gravitate [are heavy] toward one another, and [5] that there is a similar gravity [heaviness] of comets toward the sun, it will have to be concluded by this third rule that all bodies gravitate toward one another.

(ibid; bracketed numbers added)

The One-Set Reading of Rule 3 is consistent with this explanation, if we assume that Newton has adopted a novel, non-medieval interpretation of *intendi & remitti*. Specifically, the above sentence could be read as a claim that Rule 3 can be applied to gravity so long as experimental and observational evidence shows that it is a quality *both* that belongs to all bodies on which experiments and observations can be made *and* that cannot be intended and remitted insofar as its measure always varies directly with mass (cf. Section 2.3). However, such an interpretive move isn't necessary. We can instead accept the Two-Set Reading of the Rule and make equally good sense of Newton's remarks without having to assume that he is departing from the medieval usage of *intendi & remitti*.[48] Taking this tack, Newton is claiming that for Rule 3 to be applied to the gravity that he's investigating in the *Principia*, namely, a force whose measure always varies directly with mass, evidence must show that this force of gravity is among those qualities that belong to all bodies on which experiments and observations can be made. Indeed, if we assume that Newton maintains the medieval usage of *intendi & remitti*, it makes sense that evidence would have to show that gravity belongs to this particular set of universalizable qualities. For, according to standard medieval usage, gravity is a quality of bodies that can be intended and remitted insofar as the specific measure of a body's gravity will decrease and increase as the body becomes more or less distant from the body to which it is attracted.

From this Two-Set Reading, we also get a different way of understanding the distinction that Newton draws between gravity and magnetism in Corollary 5 of III.6. On this reading, both the force of gravity in one and the same body and also "the magnetic force in one and the same body can be intended and remitted" *(ibid,* 810). What makes them "of a different kind" is the general measure that is assigned to each force. The general measure of gravity always varies directly with mass, for instance, whereas

magnetic attraction is **not** proportional to the [quantity of] matter attracted. Some bodies are attracted [by a magnet] more [than in

---

48 For textual and historical evidence that Newton maintained the standard medieval use of *intendi & remitti*, see the works referenced in Note 24.

proportion to their quantity of matter], and others less, while most bodies are not attracted [by a magnet at all].

(*ibid*; boldface added)

Additionally, "as far as [Newton has] been able to tell from certain rough observations," the specific measure of magnetic force in one and the same body

> is sometimes far greater in proportion to the quantity of matter than the force of gravity; and this force, in receding from the magnet, decreases not as the square but almost as the cube of the distance ...

(*ibid*)[49]

According to Newton's report in Corollary 5 of III.6, the measure of magnetism is not always directly proportional to mass, and it does not always vary inversely to the square of the distance between a magnetic body and an attracted body. This is why the magnetic force should be taken to be "of a different kind" from the force of gravity. While both forces can be intended and remitted – insofar as the specific measures of both forces can vary – the general measure of magnetism is not characterized by the same invariant proportions that characterize the general measure of gravity.

There is another noteworthy difference between the two forces. Newton has found that "most bodies are not attracted [by a magnet at all]," which means that there is no evidentiary basis for claiming that magnetism belongs to all bodies on which experiments can be made. In turn, unlike gravity, magnetism cannot be universalized on the basis of Rule 3. For, given what Newton's "certain rough observations" have revealed, it does not belong to either set of qualities that Rule 3 identifies.

---

49 The Latin passage reads in full:

> Et vis magnetica in uno & eodem corpore intendi potest & remitti, estque non-nunquam longe major pro quantitate materiae quam vis gravitatis, & in recessu a magnete decrescit in ratione distantiae non duplicata, sed fere triplicata, quantum ex crassis quibusdam observationibus animadvertere potui.

(Newton 1871, 403)

Newton uses the basic conjunction *estque* to connect the two clauses of the sentence, which means it is not necessary to read "can be intended and remitted" as an explanation for why magnetic force is different in kind from the force gravity. This is a possible way to read the sentence and is the way that it is read by Ducheyne (2012), as noted in the paragraph preceding Note 44.

In general, then, the Two-Set Reading of Rule 3 allows us to retain the major interpretive advantages that are afforded by the One-Set Reading and to retain them without having to assume that Newton uses *intendi & remitti* in a novel, non-medieval way. As we'll see in what follows, the Two-Set Reading is also consistent with Newton's use of the Latin *quæ … quæque* construction in the statement of the Rule and his use of that construction elsewhere in the *Principia*. Rather significantly, Newton uses that construction in the first edition to signify two sets, just as I claim he is doing in Rule 3.

3.2 The *quæ … quæque* Construction in Rule 3

In both the second and third editions of the *Principia*, Rule 3 is presented as follows:

REGULA III.
   Qutalitates corprum **quæ** intendi & remitti nequeunt, **quæque** corporibus omnibus competunt in quibus experimenta instituere licet, pro qualitatibus corporum universorum habendae sunt.
(Newton 1871, 387; boldface added)

Rule 3: Those qualities of bodies **that** cannot be intended and remitted **and that** belong to all bodies on which experiments can be made should be taken as qualities of all bodies universally.
(Newton 1999, 795; boldface added)

According to the rules of Latin grammar, the *quæ … quæque* construction indicates that the subject specified in the sentence is qualified by what follows the *quæ* and also by what follows the *quæque*. However, there is an ambiguity in the meaning of Rule 3, because there are two ways to interpret the subject of the sentence. One could take the subject term "those qualities of bodies" to signify a single set of qualities, in which case Rule 3 is telling us that the single set has the features following *quæ* and also the features following *quæque*. In other words, on this One-Set Reading, the Rule identifies a single set of qualities of bodies that should be taken to belong to the general class of "qualities of all bodies universally," namely, the set that includes qualities of bodies *both* that cannot be intended and remitted *and* that belong to all bodies on which experiments can be made. Alternatively, one could take the subject term "those qualities of bodies" to signify two different sets of qualities, in which case Rule 3 is an either-or statement. It tells us that *either* those qualities of bodies that cannot be intended or remitted *or* those qualities of bodies that belong to all bodies on which experiments can be made are qualities that we ought to accept as universal qualities. That one can adopt either the One-Set or the Two-Set Reading of "those qualities of bodies" is borne out by Newton's various uses of the *quæ ... quæque* construction elsewhere in the *Principia*.

For instance, the One-Set Reading is consistent with the way Newton uses the *quæ ... quæque* construction in two sentences that appear after the section "Rules for the Study of Natural Philosophy." One sentence is in Book 3, and the other in the General Scholium. In the lead-up to the sentence that appears in Book 3, Newton presents some of the observational data of the motion of comets that had been gathered at various times and locations by people such as Montanari, Zimmerman, and Ponteo. According to Newton, the data indicates that these observers were, without their knowledge, observing the same comet. He then argues that the observed path of this single comet agrees with the theory of planetary motion that he has just presented (*ibid*, 911), and in this context, he writes:

> Et theoria, **quæ** motui tam inæquabili per maximam coeli partem probe respondet, **quæque** easdem observat leges cum theoria planetarum, & cum accuratis observationibus astronomicis accurate congruit, non potest non esse vera.
>
> (Newton 1871, 506; boldface added)

> And the theory [of comets] **that** corresponds exactly to so nonuniform a motion through the greatest part of the heavens, **and that** observes the same laws as the theory of the planets, and that agrees exactly with exact astronomical observations cannot fail to be true.
>
> (Newton 1999, 916; bracketed phrase and boldface added)

The meaning of the statement is clear. Newton is asserting that a single theory of comets "cannot fail to be true" so long as it meets all three of the conditions that he sets forth. Spelling this out, he is claiming that a particular theory of comets should be accepted as true if it *simultaneously* [1] corresponds to the observed motion of the single comet he has just identified, [2] employs the same laws as his theory of planetary motion, and, in general, [3] is consistent with the "exact astronomical observations" that are available.

Along similar lines, in a familiar passage from the General Scholium, Newton uses the *quæ ... quæque* construction to identify a series of conditions that must be met by the cause of gravity. Immediately after telling us that in the *Principia* he has "not yet assigned a cause to gravity" (*ibid*, 943), he writes:

> Oritur utique haec vis a causa aliqua, **quae** penetrat ad usque centra solis & planetarum sine virtutis diminutione; **quaeque** agit non pro quantitate *superficierium* particularum, in quas agit (ut solent causae mechanicae) sed pro quantitate materiae *solidae*; & cujus actio in immensas distantias undique extenditur, decrescendo semper in duplicata ratione distantiarum.
>
> (Newton 1871, 530; boldface added)

> Indeed, this force [of gravity] arises from some cause **that** [1] penetrates as far as the centers of the sun and the planets without any diminution of its power to act, **and that** [2] acts not in proportion to the quantity of the *surfaces* of the particles on which it acts (as mechanical causes are wont to do) but in proportion to the quantity of *solid* matter, and [3] whose action is extended everywhere to immense distances, [4] always decreasing as the squares of the distances.
>
> (Newton 1999, 943; boldface and bracketed numbers added)

As in the sentence about the theory of comets, the statement has a singular subject – the cause of gravity – and here, Newton is identifying four features that the single cause must have, given what he has established in the *Principia*. Namely, the cause of gravity must penetrate to the centers of bodies, and the action of this cause must be proportional to the mass of the bodies on which it acts, extend "everywhere to immense distances," and decrease according to the squares of the distances between bodies.[50]

However, when the subject term of a sentence that includes the *quæ ... quæque* construction is not in singular form, this sort of multiple-condition reading is not required. That is, when the subject term is in plural form, such that it represents a multiplicity, it is equally consistent with the Latin grammar to read what follows the *quæ* and what follows the *quæque* as describing different ways of classifying the subject term of the sentence.

Newton uses the *quæ ... quæque* construction in precisely this way in II.25, Theorem 19 of the first edition *Principia*:

> Corpora Funependula **quæ** in Medio quovis resistuntur in ratione momentorum temporis, **quæque** in ejusdem gravitatis specificæ

---

50 I discuss the broader relevance of Newton's statement about which features the cause of gravity must have in Section 4.2.

Medio non resistente moventur, oscillationes in Cycloide eodem tempore peragunt, & arcuum partes proportionales simul describunt.
(Newton 2009, 191; boldface added)

In line with the translation of the third edition version of this claim that is presented in Newton (1999), the theorem states:

The bobs of simple pendulums **that** [1] are resisted in any medium in the ratio of the moments of time, and **those that** [2] move in a nonresisting medium of the same specific gravity, perform oscillations in a cycloid in the same time and describe proportional parts of arcs in the same time.
(Newton 1999, 701; boldface and bracketed numbers added)

On pain of inconsistency, Newton cannot be specifying two conditions that must be met by the bobs of simple pendulums. If he were, he would be stating that the oscillations he describes would be performed by bobs that are simultaneously in a resisting and nonresisting medium. What he is actually claiming is that there are two distinct and nonoverlapping sets of pendular motion that share a particular feature. Specifically, he is claiming that the bobs of simple pendulums in *either* of these sets will belong to the general class of bobs that "perform oscillations in a cycloid in the same time and describe proportional parts of arcs in the same time."

That this is Newton's intended meaning in the first edition is manifest in the later editions of the *Principia*, where Theorem 19 of II.25 is presented as Theorem 20 of II.25, which states:

Corpora Funependula quibus, in medio quovis, resistitur in ratione momentorum termporis, & **corpora funependula** quæ in ejusdem gravitates specificæ medio non resistente moventur, oscillationes in cycloide eodem tempore peragunt, & arcuum partes proportionales simul describunt.
(Newton 1871, 295; boldface added).

In restating the theorem, Newton adds a second reference to "the bobs of simple pendulums." However, the translation of this statement that is presented by Cohen and Whitman in Newton (1999) is identical to the translation of the first edition statement, in which the term "the bobs of simple pendulums" is used just once. They do not include Newton's second reference to the bobs of single pendulums in their translation of II.25, Theorem 20, because the rules of Latin grammar do not require it. The grammatical rules also did not require Newton to revise the sentence, because, again, on pain of inconsistency, the statement of the first edition had adequately expressed that the theorem was describing two distinct and nonoverlapping sets of moving objects, one set that includes bobs moving in a resisting medium and the other that includes bobs moving in a nonresisting medium.

There is one additional instance of Newton using the *quæ ... quæque* construction in the third edition *Principia*, and like the case above, the subject term of the sentence in which it is used has a plural form. It is the second to last sentence of Book 3 and appears just before the start of the General Scholium:

> Sed fixae, **quæ** per vices apparent & evanescunt, **quæque** paulatim crescunt, & luce sua fixas tertiæ magnitudinis vix unquam superant, videntur esse generis alterius, & revolvendo partem lucidam & partem obscuram per vices ostendere.
> (Newton 1871, 526; boldface added)

> But fixed stars **that** [1] alternately appear and disappear, **and** [2] increase little by little, and [3] are hardly ever brighter than the fixed stars of the third magnitude, seem to be of another kind and, in revolving, seem to show alternately a bright side and a dark side.
> (Newton 1999, 938; boldface and bracketed numbers added)

To illuminate the ambiguity here, we can expand the remark as follows:

> But fixed stars that alternately appear and disappear, and [fixed stars that] increase little by little, and [fixed stars that] are hardly ever brighter than the fixed stars of the third magnitude, seem to be of another kind and, in revolving, seem to show alternately a bright side and a dark side.

Unlike the theorem concerning the bobs of simple pendulums, this statement, when read in isolation, does not unambiguously communicate Newton's intended meaning. It doesn't, because there is no apparent inconsistency in claiming that a set of fixed stars has all three of the features described above. A single set of fixed stars could, in principle, be such that every one of its members alternately appears and disappears, increases little by little, and appears hardly ever brighter than "the fixed stars of the third magnitude." But Newton could just as well be describing three different sets of stars and claiming that each set seems "to be of another kind."

To determine what Newton aims to communicate, we have to look at the context in which the statement is presented. And when put in context, we find that in the sentence above Newton is referring to the same fixed stars to which he referred a few lines earlier, namely, fixed stars that "can be renewed by comets falling into them and then, kindled by their new nourishment, can be taken for new stars" (Newton 1999, 937). "Of this sort," he explains, "are those fixed stars that appear all of a sudden, and that at first shine with maximum brilliance and subsequently disappear little by little" (*ibid*).[51] Newton is offering here a description of a specific and singular set of fixed stars, which suggests that his later statement is best read as a continuation of this description. That is, it suggests that in the later statement Newton is identifying three features that belong to a single set of fixed stars. Put in the terms of the sentence in which Newton uses the *quæ ... quæque* construction, he is claiming that each fixed star that can be "renewed by comets falling into [it] and ... that can be taken for [a] new" star is a star that alternately appears and disappears and that also increases little by little. What he adds at the end of this discussion is that each of these stars is also "hardly ever brighter than the fixed stars of the third magnitude."

In this instance, the *quæ ... quæque* construction leaves us with an ambiguity in meaning when the sentence is read in isolation, because the subject term of the sentence is in the plural form – "the fixed stars" – and what is described after the *quæ* is not incompatible with what is described after the *quæque*. The situation is similar with Rule 3. Its subject term is also in the plural form – "those qualities of bodies" – and the cases described after the *quæ* and after the *quæque* are not incompatible. It is in principle possible for a quality of a body to be found among all the bodies on which experiments can be made and also to be of a sort that it cannot be intended and remitted. It is also in principle possible that a quality belongs to one set but not the other. Consequently, the Latin construction of Rule 3 does not unto itself indicate whether "those qualities of bodies" refers to a single set of qualities or to two sets of qualities, as is perhaps most evident when we expand the statement of the Rule: "Those qualities of bodies that cannot be intended and remitted and [those qualities of bodies] that belong to all bodies on which experiments can be made should be taken as qualities of all bodies universally."

And so, as with the sentence about the fixed stars, we have to turn to the context in which Rule 3 is presented to determine whether the

---

51 "Hujus generis sunt stellæ fixæ, quæ subito apparent, & sub initio quam maxime splendent, & subinde paulatim evanescent" (Newton 1871, 525).

Rule is best read as describing multiple conditions that must be met by the members of a single set, or as describing multiple sets that belong to the same general class. Examining that context doesn't provide us the immediate clarity that we get in the case of the statement about the fixed stars. However, Newton does present an example of applying Rule 3 that lends more straightforward support to reading the Rule as identifying two sets of qualities. Or so I argue in the next section.

3.3 The Two Examples of Applying Rule 3

Considering first the general context in which Rule 3 appears, as situated among the four Rules for the Study of Natural Philosophy, there are two evident differences between its presentation and the presentations of Rules 1, 2, and 4. Whereas Rule 3 is followed by a lengthy two-paragraph commentary, Rules 1, 2, and 4 are each followed by no more than two sentences. Also, whereas questions linger about how best to connect the various claims that Newton makes in the commentary that accompanies Rule 3 (as I discuss more fully in Sections 4.1 and 4.2), the remarks that he appends to Rules 1, 2, and 4 clearly serve either to justify those Rules or to explain their significance.

Rule 4, for instance, directs those who are pursuing experimental philosophy to give no credence to "hypothetical" explanations that are not empirically well supported. In the remark that follows, Newton briefly explains why this Rule should be adopted: "This rule should be followed so that arguments based on induction may not be nullified by hypotheses" (Newton 1999, 796). Similarly, Rule 1 is followed by two sentences that clarify why the experimental philosopher should not multiply causes unnecessarily in her attempts to explain natural phenomena, namely, because nature operates in such a way that natural causes are not multiplied unnecessarily:

> As the philosophers say: Nature does nothing in vain, and more causes are in vain when fewer suffice. For nature is simple and does not indulge in the luxury of superfluous causes.
>
> (*ibid*, 794)[52]

---

52 Rule 1 was first presented in the second edition (1713) *Principia* as a modified version of what was Hypothesis I in the first edition (1687). The major and only difference between the two is that the commentary of Rule 1 includes the sentence that begins "As the philosophers say" (cf. Cohen 1971, 24).

Rule 2 is presented immediately thereafter, and as indicated by the "Therefore" with which it begins, this Rule also follows from the simple and economical ordering of nature:

> *Rule 2: Therefore, the causes assigned to natural effects of the same kind must be, so far as possible, the same.*
>
> Examples are the cause of respiration in man and beast, or of the falling of stones in Europe and America, or of the light of a kitchen fire and the sun, or of the reflection of light on our earth and the planets (*ibid*, 795).[53]

The examples Newton offers here aren't meant to justify Rule 2; the Rule is justified by the description of nature that precedes it. Rather, Newton is identifying the sorts of "natural effects of the same kind" for which the same causes should be assigned. In other words, with the examples he is identifying specific cases to which Rule 2 can be applied.

While there are differences between the lengthy commentary accompanying Rule 3 and the much shorter commentaries accompanying Rules 1, 2, and 4, they are not completely dissimilar. As in the commentaries of the other Rules, Newton offers a brief justification for the acceptability of Rule 3, he references the simplicity of nature, and he describes specific cases to which the Rule can be applied. The brief justification is provided in the sentence that immediately follows the statement of Rule 3, which I repeat for the sake of clarity.

> *Rule 3: Those qualities of bodies that cannot be intended and remitted and that belong to all bodies on which experiments can be made should be taken as qualities of all bodies universally [universorum].*
>
> For the qualities of bodies can be known only through experiments; and therefore qualities that square with experiments

---

53 Rule 2 was first presented in the second edition (1713) *Principia* and was there identical to what was Hypothesis II in the first edition (1687). In the second edition, the Rule stated: "Therefore[,] for natural effects of the same kind the causes are the same" (Cohen 1971, 262). The third edition version of Rule 2, quoted above, signals a shift in what Newton intended the Rule to communicate. Here he explicitly refers to the causes an experimental philosopher ought *to assign* to natural effects, not to the causes that she ought to infer *actually exist* in the natural world. For discussion of the general relevance of this shift, see Note 21.

generally [*generales*] are to be regarded as general [*generaliter*] qualities; and qualities that cannot be diminished [*minui*] cannot be taken away from bodies.

(*ibid*; bracketed Latin terms added)[54]

According to Newton, we must use experimental evidence to determine the qualities of bodies, and, broadly speaking, he's claiming that this is why the qualities identified in Rule 3 "should be taken as qualities of all bodies universally." In the one-sentence justification, he talks of "general qualities" and qualities that "cannot be taken way from bodies," but this, it seems, is just a change of terminology. With the statement providing reasons for accepting Rule 3, general qualities refer to those that belong to all bodies on which experiments can be made, and qualities that cannot be taken away from bodies refer to those that cannot be intended and remitted.[55]

The question that remains is whether Newton is claiming that evidence must indicate that a quality must be *both* a general quality *and* a quality that cannot be taken away from bodies to merit accepting it as a universal quality. This is a possible interpretation of his one-sentence justification, and it is the one that corresponds to the One-Set Reading of Rule 3. But another interpretation is possible. Newton could be claiming that to accept that a quality belongs to all bodies universally evidence must indicate that the quality is *either* a general quality *or* a quality that cannot be taken away from bodies.

---

54 "Nam qualitates corporum non nisi per experimenta innotescunt, ideoque generales statuendæ sunt quotquot cure experimentis generaliter quadrant; & quæ minui non possunt, non possunt auferri" (Newton 1871, 387–388). My translation departs from the one presented in Newton (1999). Cohen and Whitman use "universally" as the translation of "generales" and "universal" as the translation of "generaliter," whereas I use "generally" and "general," respectively.

55 In a pre-1713 draft version of Rule 3, Newton explicitly identifies qualities that be "cannot be remitted" with qualities that "cannot be taken away" from bodies. Focusing specifically on the properties of impenetrability, mobility, and inertia, he writes:

The things which cannot be intended and remitted such as ... impenetrability ... and motion ... and that inertia which causes a resistance to motion and to changes of motion ... are usually considered to be the properties of all bodies. And the reason is because **a quality which cannot be remitted cannot be taken away** ... and on the other hand that which can be taken away, if it were to be taken away from some parts of the whole, it could be remitted in the whole.

(cited in McGuire 1968, 237, with the Latin text provided on 257; boldface added)

This is the interpretation that corresponds to the Two-Set Reading of Rule 3, and it is the one that I will pursue here. Arguably, this interpretation is not uniquely picked out by Newton's commentary on the Rule. But as I show in what follows, it does illuminate why Newton provides in that commentary two very different cases to which Rule 3 can be applied and why, specifically, he claims that the Rule can be used to universalize the quality of divisibility based on the results of a single experiment.

As just noted, the fundamental point of Newton's one-sentence justification for the Rule is that since, in general, we must rely on experimental evidence to know the qualities of bodies, experimental evidence must serve as our guide when we draw inferences about which of those qualities belong to all bodies universally. On the Two-Set Reading, the more specific claim is that there are two different evidentiary circumstances in which such an inference would be justified, namely, when experimental evidence shows us that a quality of bodies "square[s] with experiments generally" and when experimental evidence shows us that a quality of bodies "cannot be diminished." To gain clarity about what these circumstances entail, and why Newton presents them as the circumstances under which it is appropriate to universalize a quality of bodies, we will examine each case in turn.

Starting with "qualities that square with experiments generally," you'll notice that, as per this basic description, such qualities are those that have generally been found to belong to all bodies on which experiments *have been made*. What Newton says in the one-sentence justification is that such qualities are "to be regarded as general qualities," which, as per the statement of Rule 3, are qualities that belong to all bodies on which experiments *can be made*. Given that this statement helps to justify the Rule, his further suggestion is that it is reasonable to infer that such general qualities are universal, that is, that they belong to all bodies universally, including those on which experiments *cannot be made*.

In making these connections, Newton is acknowledging that there are two steps of reasoning involved when we draw inferences about experimental evidence. There is an initial generalization from *some* experimental bodies to *all* experimental bodies, and then a universalizing step that proceeds from all *experimental* bodies to all bodies *universally*. He is also defending the second inference on the basis of the first. Namely, he is claiming that it is appropriate to infer that general qualities are universal qualities, because it is appropriate to regard "qualities that square with experiments generally" as general qualities, that is, to accept that qualities that have generally been found among bodies on which experiments *have been made* are the qualities of bodies on which experiments *can be made*. Newton's brief explanation for why it is appropriate to draw this sort of some-to-all inference is presented immediately after the one-sentence justification:

> Certainly idle fancies ought not to be fabricated recklessly against the evidence of experiment, nor should we depart from the analogy of nature, since nature is always simple and ever consonant with itself.
>
> (ibid)

In justifying Rules 1 and 2, Newton took for granted that any person pursuing the experimental philosophy of the *Principia* would agree that "nature is simple and does not indulge in the luxury of superfluous causes." He also took for granted that they would accept that our reasoning about natural causes and effects should remain faithful to nature's simple and economical ordering. In the case of Rule 3, he is making the same broad assumption; he is assuming that his reader will accept that nature has a simple and uniform ordering and accept as well that the inferences we draw from experimental evidence should remain faithful to the order of nature. (Of course, he is additionally assuming that the kinds of reasoning identified by these three Rules are consistent with the general order of nature and would thus be acceptable to his readers.)

This part of Newton's justification for Rule 3 illuminates the rationale behind some of the key steps of the argument for universal gravity. Drawing on experimental results produced by Huygens, Newton reasons that *all* terrestrial bodies fall according to an inverse-square force (cf. Section 2.1). Drawing on experimental results he produced, he reasons that the weights of *all* terrestrial bodies equidistant from the center of the earth are directly proportional to their masses (cf. Section 2.2). When these results are presented in the proofs for III.4 and III.6, respectively, Newton provides no justification for drawing the inference from *some* experimental bodies of a particular type to *all* experimental bodies of the same type. And, as we see from the presentation of the first three Rules, no special argument is offered in the body of the proofs, because he takes for granted that his readers accept that "nature is always simple and ever consonant with itself," and that the some-to-all reasoning that he applies in these cases is consistent with nature's simple and economical ordering. The additional point he makes in defense of Rule 3 is that it is equally acceptable, and equally unproblematic, to draw inferences from bodies on which experiments *can be made* to bodies on which experiments *cannot be made*. In other words, he is pointing out that the same store of experimental evidence that supports regarding qualities as general qualities also supports regarding those qualities as universal – as qualities belonging even to those bodies and parts of bodies that are imperceptible.

An important practical consequence follows. To accept Newton's explanation is to accept that when our natural philosophical investigations show us that a quality of bodies is found to "square with experiments generally," we can take that quality to be a quality that belongs to all bodies universally. This is the point that Newton urges as he explains the circumstances under which Rule 3 could be used to universalize gravity. We have seen his explanation a couple of times already, in Sections 2.3 and 3.1, but the sentence bears repeating:

> Finally, if it is universally established [*universaliter constet*] by experiments and astronomical observations [1] that all bodies on or near the earth gravitate [*lit.* are heavy] toward the earth, and do so in proportion to the quantity of matter in each body, and [2] that the moon gravitates [is heavy] toward the earth in proportion to the quantity of its matter, and [3] that our sea in turn gravitates [is heavy] toward the moon, and [4] that all the planets gravitate [are heavy] toward one another, and [5] that there is a similar gravity [heaviness] of comets toward the sun, **it will have to be concluded by this third rule that all bodies gravitate toward one another**.
>
> (*ibid*, 796; boldface, bracketed numbers,
> and Latin phrase added)

In the situation Newton describes, gravity has been discovered to be a quality that squares with a wide store of experimental and observational evidence. It is a situation in which we thus have at our disposal the empirical evidence that we need to regard gravity as a general quality, that is, to draw the (allegedly) unproblematic inference that, as a quality that is generally found among all bodies on which experiments and observations *have been made*, gravity is also a quality that belongs to all bodies on which experiments and observations *can be made*. This is a kind of quality that Rule 3 tells us is universalizable, and so, applying the Rule, we can take another (allegedly) unproblematic step in our reasoning and regard gravity as a universal quality, that is, as a quality that also belongs to bodies on which experiments *cannot be made*.

Newton's other claim in the one-sentence justification for Rule 3 is connected with a different form of reasoning that can be applied to experimental evidence. The basic claim there, recall, is that when experimental evidence shows us that a quality of bodies "cannot be diminished," we can legitimately infer that the quality is a quality of all bodies universally. What Newton specifically says is that qualities of bodies that "cannot be diminished" are qualities that "cannot be taken away from bodies" (*ibid*, 795). Insofar as this remark helps justify Rule 3, his further claim, as noted above, is that qualities that "cannot be taken away from bodies" refer to qualities that cannot be intended and remitted, and thus to qualities that Rule 3 identifies as universalizable.[56] The question that remains is why Newton draws these connections. What assumptions is he making about the appropriate ways to reason about experimental evidence when he claims that those qualities that evidence indicates cannot be diminished are qualities that cannot be taken away from bodies, and when he additionally claims that such enduring qualities are to be counted as qualities that cannot be intended and remitted? There also remains the related question of why the universalizing inference that is sanctioned by Rule 3 would be needed in a case where an experimental philosopher has adequate evidence to claim that a quality cannot be taken away from bodies. It would seem that if she has experimentally established that a quality is an enduring quality of bodies, she has also and already established that the quality belongs to all bodies universally.

For answers to these questions, we can turn to Newton's additional example of what follows from applying Rule 3. The example ends the first paragraph of the commentary that accompanies the Rule and immediately precedes the example of universalizing gravity. Newton writes:

> Further, from phenomena we know that the divided, contiguous parts of bodies can be separated from one another, and from mathematics it is certain that the undivided parts can be distinguished into smaller parts by our reason. But it is uncertain whether those parts which have been distinguished in this way and not yet divided can actually be divided and separated from one another by the forces of nature. But if it were established by even a single experiment that in the breaking of a hard and solid body, any undivided particle underwent division, we should conclude by the force of this third rule not only that divided parts are separable but also that undivided parts can be divided indefinitely.
>
> (*ibid*, 796)

---

56 Manuscript evidence that Newton identifies qualities that "cannot be taken away from bodies" with qualities that "cannot be intended and remitted" is supplied in Note 55.

Newton describes a situation in which there is an evident difference between what can be concluded on the basis of mathematical reasoning and what can be concluded based on experimental evidence. The mathematician concludes with certainty that the undivided parts of a body can be divided. And she does, because, as Newton describes, she thinks about the parts of a body that have not yet actually been divided, and she distinguishes such parts by use of reason. The question remains as to whether a body that has been rationally divided in this way could also and in fact be divided "by the forces of nature." This is a question that must be settled based on empirical evidence, and thus, it is a question for an experimental philosopher, not a mathematician.[57]

According to Newton, the experimental philosopher need perform only a single experiment to draw a positive conclusion. She must break just one "undivided particle" of "a hard and solid body" and experimentally confirm that one part of a body that previously could not be divided by experiment can be divided. If she succeeds, then, on Newton's account, she would have sufficient evidence to apply Rule 3 and universalize the quality of divisibility. Namely, by Rule 3 she could conclude "that undivided parts can be divided indefinitely," that is, she can accept that all of the not-yet divided parts of all bodies can be continuously divided.[58] Newton notes that by Rule 3 the experimental philosopher can also conclude "that divided parts are separable." However, this inference is not based on the results of the single experiment. Rather, the Rule can be used to universalize the quality of separability, because evidence indicates that this is a *general* quality - or, as he puts it in the first sentence of the example, because "**from phenomena** we know that the divided, contiguous parts of bodies can be separated from one another" (*ibid*; boldface added).

---

57 For discussion of the role played by mathematical certainty in the method of the *Principia*, see Guicciardini (2009) and Domski (2018).

58 Although Newton casts the results of the single experiment in terms of what can be asserted of all bodies and all their parts, it is possible to view the significance of the example in negative terms. Specifically, as pointed out to me by Zvi Biener, one might take Newton to be describing the sort of experimental evidence that could disprove the existence of fundamentally indivisible parts of matter and thereby serve as a counterexample to atomism. Given that a commitment to atomism appears to be part and parcel of Newton's experimental philosophy (as I discuss in Section 4.1), one might additionally take Newton to be suggesting that actually producing this sort of counterexample is experimentally impossible. Such a reading is compatible with the presentation of the example that I provide, though it is not necessary to identify this as Newton's broader intention to make sense of the role of Rule 3 in the example.

On what grounds, then, is Rule 3 applied to the single experiment that Newton describes? Given the details of the example, this is not a situation where the experimental philosopher has identified a quality that "square[s] with experiments **generally**" (ibid, 795; boldface added). She has identified a quality that squares with the single experiment that she has conducted, and thus, as per the criterion that Newton explicitly sets forth in the one-sentence justification for Rule 3, she does not have the evidence required to claim that divisibility is a general quality, that is, to claim that it is a quality that belongs to all bodies on which experiments can be made. This leaves us with the other option Newton presents: Rule 3 is applied to the results of the experiment, because the experiment indicates that the body that has been divided has a quality that "cannot be diminished." As per Newton's one-sentence justification for the Rule, the experimental philosopher has thereby identified a quality that "cannot be taken away from bodies" and, in turn, a quality of bodies that cannot be intended and remitted.

That divisibility is a quality that "cannot be diminished" – that it is a quality not subject to decrease – seems straightforward enough, since, by its basic definition, divisibility does not vary in degree. This is a quality for which the question "how much?" does not make sense. Either a body (or a part of a body) will be divisible or it will not. Even if we continue to divide the parts of a body that's been divided, for each part that we divide there isn't more or less divisibility. We could start with some piece of steel, for instance, and divide the piece in half and then half the two halves. We may even continue this process until we've produced pieces that are too small for us to divide by any experimental means that we have available. But, for each piece that we have divided, there is not more or less divisibility. The quality of divisibility simply belongs to each piece that's been divided.

What is less obvious is why having experimental confirmation that divisibility belongs to one part of a single body is adequate for claiming that divisibility belongs to some other body, let alone for claiming that it "cannot be taken away bodies." It is less obvious because, as in the example of universalizing gravity, Newton takes for granted the steps of reasoning that are needed to generalize what has been established by the experiment. In the case of gravity, there was an initial generalization that proceeded from *some* experimental bodies to *all* experimental bodies, that is, from bodies on which experiments and observations *have been* made to bodies on which experiments and observations *can be* made. Only after this conclusion has been drawn can Rule 3 be applied. For Rule 3 to be applied to divisibility, the experimental philosopher must instead reason from her experimental discovery that divisibility belongs to a *single sensible part* of a sensible body to the conclusion that it belongs to *all the sensible parts* of that body, and then, on that basis, infer that the same quality belongs to *all the sensible parts of all sensible bodies*. In other words, from her experimental result she reasons that divisibility cannot be taken away from any sensible part of any sensible body and, in turn, that divisibility is a quality of all sensible bodies that cannot be intended and remitted. Applying Rule 3, she can then universalize the quality and accept that divisibility belongs to all sensible and insensible bodies and to all the sensible and insensible parts of all bodies.

That this specific sort of inference is at stake in the example of divisibility – that it is meant to illustrate a progression from what is sensible to what is insensible – is signaled by the discussion that precedes the example, where Newton describes the part-to-whole reasoning that is to be adopted by the experimental philosopher. In this context, Newton specifically indicates that pursuing experimental philosophy requires a commitment to the two basic tenets of atomism, namely, that sensible bodies are composed of insensible parts and that the qualities of a sensible body originate from the qualities of a body's insensible parts (cf. Newton 1999, 795–796). In Section 4.1, we will turn to the question of whether Newton is additionally claiming that the commitment to these tenets must be based on knowledge that natural bodies have this sort of atomistic structure. For now, notice that the example of divisibility provides an illustration of how these commitments can be applied to experimental evidence. Specifically, the atomist part-whole characterization of how natural bodies are structured provides the experimental philosopher a basis for reasoning from the evidence she has gathered from a single sensible part of a body to all the body's sensible parts. Coupling nature's general uniformity and simplicity with the related atomist tenet that all natural bodies share the same part-whole structure, she can also reasonably infer that a quality found among all the sensible parts of one sensible body belongs to all the sensible parts of all sensible bodies. Applying Rule 3 then permits her to extend this conclusion and accept that all bodies universally, including those bodies and parts of bodies that are insensible, have the very same quality that belongs to all sensible bodies and all their sensible parts.

The way that Newton puts it in the example of divisibility, the experimental philosopher who has divided one "undivided particle" of one "hard and solid body" can "conclude by the force of [the] third rule" that the quality of divisibility belongs to all the undivided parts of a body, that is, to all the parts of a body that have yet to be divided. In practice, this means that however far the experimental philosopher is able to divide the parts of a body by experimental means, she can accept that the parts she has left undivided could be divided. It also means that if those not-yet-divided parts were divided, such that a new set of not-yet-divided parts was produced, she can accept that those newly produced undivided parts are also divisible. Indeed, what the application of Rule 3 allows her to infer is that no matter how far the process of division might continue, any and all not-yet-divided parts that are produced – including those that are too small for her to sense – are divisible. In Newton's terms, the application of the Rule allows her to accept "that undivided parts can be divided **indefinitely**" (*ibid*, 796; boldface added).

Tracking the reasoning that is at play in the example puts us in a better position to understand the significance of Newton's claim that qualities of bodies that "cannot be diminished" are qualities that "cannot be taken away from bodies." Namely, the example of divisibility indicates that there are two generalizations that inform the connection that he draws. If evidence indicates that some sensible part of a body has a quality that cannot be decreased – or, more generally, has a quality that does not vary in degree – we first generalize that quality to every other sensible part of the body. We then make a further generalization and infer that the quality belongs to all the sensible parts of all bodies. It is in this particular respect that the quality "cannot be taken away from bodies." It is a quality that belongs to all the *sensible* parts of bodies and belongs to them in such a way that it cannot be diminished. Moreover, since the qualities of a whole body are taken to arise from the qualities of its parts, divisibility is a quality that will also be retained by *whole* sensible bodies in such a way that it will not vary in degree, that is, in such a way that its specific measure in the whole body will not increase or decrease. Consequently, it is quality that experimental evidence indicates is a quality of all sensible bodies and one that, for all such bodies, cannot be intended and remitted.

We now also have an explanation for why the universalizing inference sanctioned by Rule 3 is to be applied to qualities that evidence has shown are qualities that "cannot be taken away from bodies." As we've just seen, these are not qualities that Newton presents as enduring in all bodies in general. They are qualities that are taken to endure in all *sensible* bodies and all their *sensible* parts. Rule 3 is then applied to widen their scope. That is, the Rule is applied so that we can take these qualities to be qualities of all bodies universally, including those parts of bodies that are insensible.

At this juncture, we also have a clearer view of how Newton's two examples of applying Rule 3 are connected. Perhaps most evidently, in both the example of gravity and the example of divisibility, the application of the Rule yields the same result; it allows us to draw inferences about the qualities that belong to bodies that are out of our experimental and sensory reach. Whether we characterize such bodies as bodies on which experiments cannot be made or as bodies that are insensible, the Rule allows us to extend to these bodies a feature that has been generalized to all experimental and sensible bodies. And here, in regard to the reasoning that allows us to apply Rule 3, there is another similarity between the examples. Both highlight that reasoning from a particular set of evidence to a general conclusion – and specifically, from one or some bodies to all the bodies in the range of our experiments and senses – rests on a commitment to the simplicity and uniformity of nature. More precisely, such generalizations are grounded, for Newton, on two basic premises: That "nature is always simple and ever consonant with itself," and that we should not "depart from the analogy of nature" when reasoning about the qualities of bodies (*ibid*, 795). Both examples exemplify the importance of these commitments. In the case of gravity, nature's simple and uniform ordering supports the inferences that we draw from bodies on which experiments *have been* made to bodies on which experiments *can be* made. Just the same, in the case of divisibility, the simplicity and uniformity of nature lend support to the inferences that we draw from the sensible parts of *one* sensible body to the sensible parts of *all* sensible bodies.

Of course, a commitment to nature's simplicity and uniformity is not enough. As both examples also make clear, experimental and empirical evidence plays a fundamental role when the experimental philosopher universalizes a quality of bodies on the basis of Rule 3. And on the Two-Set Reading that we have been pursuing, it is the evidence that's used in the examples that clarifies why Newton presents two very different cases of applying the Rule. In the case of gravity, there is evidence that this force generally belongs to all bodies on which experiments and observations have been made, and therefore evidence that it is a general quality of bodies. In the case of divisibility, there is evidence that this quality, which cannot be diminished, belongs to one sensible part of one body, and therefore evidence that it cannot be taken away from whole sensible bodies. While in both cases the evidence supports a generalization that extends to all experimental and sensible bodies, the evidence in the examples supports the generalization for different reasons. In one case, the evidence indicates that a quality is a general quality; in the other, it indicates that the quality cannot be intended and remitted. Consequently, by providing two different examples, Newton is clarifying why two different evidentiary circumstances support the universalization that is sanctioned by Rule 3.

The One-Set Reading of Rule 3 gives us a different way of connecting Newton's two examples, and it requires taking a different approach to Newton's explicit remarks. On that reading, you'll recall, the Rule expresses the two necessary and sufficient conditions that are to be met by universalizable qualities. This means that the experimental and empirical evidence that Newton presents in both of the examples of applying the Rule must somehow show that the qualities being universalized – gravity in one case, divisibility in the other – are *both* general qualities *and* qualities that they cannot be intended and remitted. We saw in Section 2.3 how this reading applies to the example of gravity. The experimental and observation evidence to which Newton refers is evidence that gravity is a general quality. It is also evidence that gravity cannot be intended and remitted insofar as the general measure of gravity is characterized by invariant proportions. Newton does not claim that he is using *intendi & remitti* in this non-medieval way, but, as we saw in Section 2.3, on the One-Set Reading, it seems his use of the phrase has to be understood along these lines to make sense of the Rule's application to the proportionality between weight and mass in Corollary 2 of III.6.

A similar approach would have to be taken with the example of divisibility. In this instance, what's specifically at issue is why the evidence that's been gathered by the experimental philosopher establishes that divisibility is a general quality. It is at issue, because Newton's explicit characterization of general qualities doesn't fit the circumstances of the example. In the example, the experimental philosopher has found divisibility to belong to a single experimental body, and what Newton states in the one-sentence justification for Rule 3 is that "qualities that square with experiments generally are to be regarded as general qualities" (*ibid*, 795). Consequently, to explain why the experimental philosopher has adequate evidence to infer that divisibility is a quality of bodies that belongs to all bodies on which experiments can be made – and thus, adequate evidence to apply Rule 3 – we would need to go beyond Newton's explicit remarks and posit some additional criterion for "general qualities" that is being used in the example.

Adopting the Two-Set Reading of Rule 3, we don't need to make such interpretive moves to determine which qualities count as general qualities or which count as those that cannot be intended or remitted. On the Two-Set Reading, a quality can be universalized if it is *either* a general quality *or* a quality of bodies that cannot be intended and remitted, and following Newton's one-sentence justification for Rule 3, a general quality just is a quality that "square[s] with experiments generally" and a quality of bodies that cannot be intended and remitted just is a quality that "cannot be diminished." What the example of applying Rule 3 to divisibility illustrates is that evidence need only show that a quality that does not vary in degree belongs to a single sensible part of a sensible body for us to regard it as a quality that "cannot be taken away from bodies" – as a quality of all sensible bodies that cannot be intended and remitted – and thus, to regard it as a type of quality that Rule 3 tells us can be universalized. The example of applying Rule 3 to gravity illustrates that the Rule can also be applied under different evidentiary circumstances. When there is evidence that a quality is generally found among those bodies on which experiments and observations have been made, as per Newton's characterization, we have evidence that the quality is a general quality, and thus, as per Rule 3, one that should be taken to belong to all bodies universally.

On the Two-Set Reading, then, we can make sense of Newton's two examples in a more straightforward way. We need not speculate about what nuanced way Newton is using *intendi & remitti* or "general qualities," as one does if they adopt the One-Set Reading. According to the Two-Set Reading, there are two different sets of qualities that Rule 3 tells us can be universalized, and on this reading, the two examples of applying the Rule illustrate the point suggested by the one-sentence justification for the Rule that opens the accompanying commentary, namely, that the evidence that's needed to determine membership in these different sets is of a different kind.

This brings us to a further advantage of the Two-Set Reading. On this reading Newton has the resources to justify his claim that gravity and inertia should be regarded as universal forces of a different kind. I spell out this advantage in Section 4.2. Immediately below, in Section 4.1, I set the stage for that discussion by taking a closer look at the rhetoric that characterizes Newton's justification for Rule 3 and the role of atomism in his program of experimental philosophy.

4 Universal Qualities and Explaining the Phenomena

4.1 Atomism in the Argument for Rule 3

The four Rules for the Study of Natural Philosophy highlight the importance of empirical evidence for the program of experimental philosophy that Newton pursues in Book 3 of the *Principia*. The first three Rules identify the kind of empirical evidence that is needed to draw inferences about natural causes and effects and about the qualities that belong to all natural bodies. The fourth Rule tells us that only empirical evidence – "yet other phenomena," as Newton puts it – is to be used to revise propositions that have been "gathered from phenomena by induction" (Newton 1999, 796).

The way in which Newton presents the Rules also illuminates the specific sense in which the reasoning used in experimental philosophy is non-hypothetical. In the General Scholium, Newton makes the broad claim that "whatever is not deduced from the phenomena must be called a hypothesis; and hypotheses, whether metaphysical or physical, or based on occult qualities, or mechanical, have no place in experimental philosophy" (*ibid*, 943). With Rule 4, Newton clarifies that hypotheses specifically have no place when evaluating the status of propositions that have been established based on evidence and induction. According to this Rule, "In experimental philosophy, propositions gathered from phenomena by induction should be considered either exactly or very nearly true **notwithstanding any contrary hypotheses**, until yet other phenomena make such propositions either more exact or liable to exceptions" (*ibid*, 796; boldface added). And this Rule should be followed, Newton says, precisely so that "arguments based on induction may not be nullified by hypotheses" (*ibid*). The rejection of hypotheses also comes into play as Newton explains why the experimental philosopher ought to accept Rule 3. In the commentary that accompanies that Rule, he notes that "idle fancies ought not

DOI: 10.4324/9781003184256-4

to be fabricated recklessly against the evidence of experiment" (*ibid*, 795). Or, put in the terms of the Motte-Cajori (1934) translation of the *Principia*, reasoning appropriately about empirical evidence requires giving no consideration to "dreams and vain fictions of our own devising" (Newton 1934, Vol. II, 398).[59]

As we have seen, the experimental philosopher instead accepts that reasoning appropriately about empirical evidence requires not "depart[ing] from the analogy of nature" (Newton 1999, 795). For her, "Nature does nothing in vain," "nature is simple and does not indulge in the luxury of superfluous causes," and "nature is always simple and ever consonant with itself" (Newton 1999, 794–795). And when considering evidence that has been gathered from nature, she will reason in a way that remains faithful to this simple, economical, and uniform order of nature. As Newton emphasizes in the commentary to the first three Rules, it is because of a commitment to this non-hypothetical manner of reasoning that the experimental philosopher will specifically accept that she should not multiply causes unnecessarily, that she should assign similar causes to similar effects, and that she should take qualities of bodies that cannot be intended and remitted and qualities that belong to all bodies on which experiments can be made to be the qualities of all bodies universally.

Newton's presentation of the Rules thus makes clear that the non-hypothetical manner in which the experimental philosopher reasons about empirical evidence is premised on a commitment to a portrait of nature in which simplicity, economy, and uniformity characterize

---

59 "Certe contra experimentorum tenorem somnia temere confingenda non sunt, nec a naturæ analogia recedendum est, cum ea simplex esse soleat & sibi semper consona" (Newton 1871, 388). In the "Editor's Preface to the Second Edition," Roger Cotes singles out Descartes's Vortex Hypothesis as one of the "dreams," or "idle fancies" (*somnia*), that should be excluded as an explanation of natural phenomena. Cotes writes that when natural philosophers "take the liberty of imagining that the unknown shapes and sizes of particles are whatever they please, and of assuming their uncertain positions and motions, and even further of feigning certain occult fluids that permeate the pores of bodies very freely, since they are endowed with an omnipotent subtlety and are acted on by occult motions: when they do this, they are drifting off into dreams, ignoring the true constitution of things [*jam ad somnia delabuntur, neglecta rerum constitutione vera*], which is obviously to be sought in vain from false conjectures, when it can scarcely be found out even by the most certain observations. Those who take the foundation of their speculations from hypotheses, even if they then proceed most rigorously according to mechanical laws, are merely putting together a romance [*fabulam*], elegant perhaps and charming, but nevertheless a romance [*fabulam*]" (Newton 1999, 385–386; bracketed Latin terms added).

nature's order and operations. However, the particular status that he grants to this general portrait of nature at the opening of Book 3 is importantly different than the status Descartes grants to the principles of metaphysics and physics that are foundational to the natural philosophy of the *Principles*.

We saw in Section 1.1 that Descartes's natural philosophy rests on three fundamental principles, namely, [1] that the soul exists; [2] that God exists as the "author of everything which is in the world" and as "the source of all truth," and [3] that there exist bodies that are "extended in length, width, and depth, which have diverse figures and are moved in diverse ways" (Descartes 1984, xxii). These, according to Descartes, are "among the most evident and most clear [things] which the human mind can know," and they are also "such that one can deduce from them the knowledge of all other things which are in the world" (*ibid*, xxi–xxii). In the *Principles*, Descartes aims to convince his readers of both of these points. He proves in Parts I and II that these principles can be evidently and clearly known by any reader who employs her reasoning in the appropriate way.[60] He then shows in the later portions of Part II that additional truths about nature can be deduced from our understanding of God's nature and our awareness that bodies as essentially extended.

In the *Principia*, Newton takes on neither of these projects. He nowhere alleges that from nature's simplicity, economy, and uniformity one can establish what is certainly and indubitably true of the natural world. Indeed, with his statement of Rule 4, he openly admits that in experimental philosophy the possibility always lingers that the conclusions we draw when our reasoning is guided by a commitment to nature's simple, economical, and uniform ordering could turn out to be false, because the possibility always lingers that additional evidence will be found that indicates that our empirically grounded proposals must be revised. Accordingly, Newton does not fashion the "deductions" from the phenomena that are used in experimental philosophy as deductions that yield absolute truths about nature. They are arguments that are stronger or weaker depending on the store of evidence on which they are based.

Newton makes this point explicit after presenting the example of applying Rule 3 to gravity, as he compares the "arguments from

---

60 Descartes demonstrates the existence of the soul in I.7–9, presents proofs for the existence of God in I.14 and I.17–21, and argues that the nature of body consists in extension in II.3–4.

phenomena" by which the qualities of gravity and impenetrability can be generalized. You'll recall that, according to Newton's example, gravity could be universalized by Rule 3 if there was, at minimum, adequate evidence to consider it a general quality of bodies, namely, if the observations made by astronomers provided empirical confirmation that gravity can be found among celestial bodies such as the planets, the earth's moon, and the comets, and also if there was experimental evidence that gravity can generally be found among all terrestrial bodies including the tides (Newton 1999, 796).[61] A similar, evidence-based argument could also be applied to impenetrability, that is, to the quality of bodies by which they fill a space such that other bodies are unable to pass through the space that they occupy. As a quality that has generally been found among all the terrestrial bodies on which experiments *have been made*, impenetrability merits being regarded as a general quality, that is, as a quality that belongs to all bodies on which experiments *can be made*. However, according to Newton, such an argument for generalizing impenetrability will remain weaker than the argument for generalizing gravity, because in contrast to gravity, there is no experimental or observational evidence confirming that impenetrability belongs to a celestial body. As Newton puts it:

> Indeed, the argument from phenomena will be even stronger for universal gravity than for the impenetrability of bodies, for which, of course, we have not a single experiment, and not even an observation, in the case of heavenly bodies.
>
> (*ibid*)

In addition to not claiming that arguments from phenomena can yield the certainty, or indubitability, that Descartes ascribes to the conclusions that he deduces from his first principles of metaphysics and physics, Newton also does not present in the *Principia* a direct argument to support the image of nature that is fundamental to his experimental philosophy. He nowhere attempts to demonstrate that nature essentially and in fact "does nothing in vain," that it "is simple and does not indulge in the luxury of superfluous causes," or that it "is always simple and ever consonant with itself" (*ibid*, 794–795). This, "As the philosophers say," is the way the natural world is ordered, and

---

61 I say "at minimum," because on the One-Set Reading of Newton's example, there would also have to be evidence that gravity cannot be intended and remitted (cf. Section 2.3).

Newton shows that this general and generally accepted picture of nature points to specific directives about the appropriate ways to reason about empirical and experimental evidence. In this respect, Newton uses a method of persuasion that is markedly different than the one used by Descartes. In the *Principles*, Descartes identifies for his readers the steps of reasoning they are to take to become convinced that the foundational principles of his natural philosophy are indubitable. Newton instead begins from the assumption that his readers accept the basic account of nature on which experimental philosophy rests, and he takes for granted that, for them, it is indubitable that nature is simply, economically, and uniformly ordered. To then persuade these readers to adopt the Rules in their investigations of nature, he shows that these directives for how to reason about empirical evidence are consistent with and supported by the image of nature to which they are already committed.[62]

Newton uses the same strategy when he explains why it is acceptable to make the universalizing inference that Rule 3 warrants – an inference that allows us to extend qualities that we have found among bodies we have sensed *both* to sensible bodies that we have yet to encounter *and* to insensible bodies that are impossible for us to perceive. We saw in the previous chapter (cf. Section 3.3) that Newton opens the commentary of Rule 3 with a one-sentence justification for the Rule in which he specifies the sorts of experimental evidence that indicate that a quality of bodies is universalizable. According to that opening sentence, evidence that a quality of bodies "square[s] with experiments generally" is evidence that it should be regarded as a "general quality," and evidence that a quality "cannot be diminished" is evidence that the quality "cannot be taken away from bodies" (*ibid*, 795). Newton then cautions that when reasoning about experimental evidence, we should refrain from positing "idle fancies" and also not "depart from

---

62   There are other texts in which Newton does present a direct argument for the superiority of his manner of investigating nature. The pre-*Principia* manuscript *De Gravitatione* is a notable example. In the opening portion of that tract, Newton presents a series of arguments that are meant to demonstrate that Descartes has provided in the *Principles* a mistaken account of the relationship between space and body and, in turn, that Descartes has not provided an adequate treatment of natural motions. Newton then presents alternative conceptions of space and body, and in a manner not unlike the one Descartes uses in the *Principles*, Newton defends his account of body by showing that it is consistent with our limited understanding of how God created the natural world. The now-standard translation of *De Gravitatione* was produced by Christian Johnson with assistance from Andrew Janiak and can be found in Newton (2004).

the analogy of nature" (*ibid*). In the next six lines, he illustrates what it means to reason on the assumption that "nature is always simple and ever consonant with itself" by describing the reasoning that we use when we extend the qualities of extension, hardness, impenetrability, mobility, and inertia to all bodies, including those bodies that are imperceptible. Newton writes:

> The extension of bodies is known to us only through our senses, and yet there are bodies beyond the range of these senses; but because extension is found in all sensible bodies, it is ascribed to all bodies universally. We know by experience that some bodies are hard. Moreover, because the hardness of the whole arises from [*oritur*] the hardness of its parts, we justly infer from this not only the hardness of the undivided particles of bodies that are accessible to our senses, but also of all other bodies. That all bodies are impenetrable we gather not by reason but by our senses. We find those bodies that we handle to be impenetrable, and hence we conclude that impenetrability is a property of all bodies universally. That all bodies are movable and persevere in motion or in rest by means of certain forces (which we call forces of inertia) we infer from finding these properties in the bodies that we have seen.
>
> (*ibid*; bracketed Latin term added)

What is notable about Newton's remarks is that he is not offering directives for how to engage with empirical evidence when doing natural philosophy. He is pointing to the reasoning that we use in our everyday encounters with natural objects and pointing out that the inferences we standardly draw are ones that are based on what is disclosed to us by our senses. He notes, for instance, that we are not convinced that all bodies are extended because this is something that is discovered by use of our reason. Rather, we ascribe extension to all bodies universally – to all bodies we have yet to sense and to those "beyond the range of [our] senses" – because we have found that extension belongs to all the bodies that we have sensed.

Newton additionally points out that the inferences that we commonly draw from some sensible bodies to all bodies universally are based on a commitment to a particular image of nature. Broadly speaking, it is an image according to which "nature is always simple and ever consonant with itself." But, more specifically, it is an image according to which there are qualities belonging to the insensible parts of a body that give rise to the same qualities as they are sensed in the whole body. We would not "justly infer," for instance, that the insensible parts of a hard body are hard if we did not also and already accept that "the hardness of the whole arises from the hardness of its parts." According to Newton, we would not make any of our standard inferences about the universality of hardness, extension, impenetrability, mobility, and inertia –we would not infer that these are qualities that belong to all the imperceptible parts of all bodies – if we did not also and already accept that the reason we perceive whole bodies to have these qualities is because all the parts of bodies have these very same qualities. Immediately after the six lines quoted above, he explains:

> The extension, hardness, impenetrability, mobility, and force of inertia **of the whole arise from** [*oritur*] the extension, hardness, impenetrability, mobility, and force of inertia **of each of the parts**; **and thus we conclude** that every one of the least parts of all bodies is extended, hard, impenetrable, movable, and endowed with a force of inertia. And this is the foundation of all natural philosophy.
>
> (795–796; bracketed Latin term and boldface added)

Read in light of the general rhetoric that he uses when presenting the Rules, Newton is indicating here that the "foundation of all natural philosophy" (*fundamentum philosophiæ totius*) is none other than a principle already accepted by those who do not question that nature operates simply, economically, and uniformly, and who additionally accept that the qualities of bodies can only be known through empirical evidence. In other words, he is emphasizing that the foundational principle of experimental philosophy is the very same principle that informs the inferences such readers make when they claim that extension, hardness, impenetrability, mobility, and inertia are qualities of all natural bodies universally. More generally, he is showing these readers that the atomist image of nature that underwrites the reasoning they standardly use when considering the relationship between a selection of qualities found among whole bodies and the qualities belonging to the parts of these bodies is the very same image of nature on which Rule 3 is based and also the very same image of nature that is foundational to the experimental philosophy that is pursued in Book 3 of the *Principia*.[63]

---

63 In draft materials from the period before Rule 3 first appeared in print, Newton says Rule 3 "seems to be the foundation of all Philosophy. For otherwise one could not derive the qualities of imperceptible bodies from the qualities of perceptible (bodies)" (cited in Cohen 1966, 175–176). On the reading I have just offered, Newton opted in the second and third editions of the *Principia* to instead identify "the

Remaining sensitive to the rhetoric of Newton's remarks, the commentary that accompanies Rule 3 does not serve to justify the Rule or the "foundation of all natural philosophy" in a traditional sense. He argues for Rule 3 by showing that what allegedly follows from the two fundamental premises of his experimental philosophy – that knowledge of bodies is gained only through experimental and empirical evidence and that "nature is always simple and ever consonant with itself" – are the same principles of reasoning that are applied by the readers of the *Principia* who accept these premises. But he does not argue for their general acceptability. In the commentary of Rule 3, he provides no justification for the claim that the qualities of bodies can only be known through experimental and empirical evidence, and he supplies no evidence to convince us that nature is simply, economically, and uniformly ordered. He also spells out no argument that takes us from nature's simple, economical, and uniform operations to the atomist commitments that are foundational to experimental philosophy. It is up to his readers to notice and accept that the two are connected.[64]

---

foundation of all natural philosophy" with the part-whole account of qualities that is applied when the qualities of extension, hardness, impenetrability, mobility, and inertia are universalized. For an alternative interpretation, see Cohen (1966, 1971), Mandelbaum (1964), McGuire (1968), and Okruhlik (1989), all of whom indicate that Newton presents Rule 3 as the "foundation of all natural philosophy" in the second and third editions of the *Principia*. My reading of the foundational role that Newton assigns the atomist account of qualities is consistent with the one offered by Belkind (2017), who claims that, for Newton, "the atomist thesis becomes a methodological requirement" (Belkind 2017, 680). Belkind additionally defends a novel interpretation of "qualities of bodies that cannot be intended and remitted" that stems from Newton's "methodological atomism" (*ibid*, 680). According to Belkind's interpretation, a quality that cannot be intended and remitted is, for Newton, one that is "invariant under changes in the configuration of the atomic parts" (*ibid*, 677). Belkind also argues that gravity is such a quality (*ibid*, 691–696). I part ways with Belkind on these specific points, because, as explained in Chapter 3, I read Newton as retaining the standard medieval usage of *intendi & remitti* and as accepting gravity to be a quality that can be intended and remitted.

64 My emphasis here is different, though not inconsistent, with the emphasis of McGuire (1970) and McMullin (1978). Whereas I focus on the commitments that Newton requires for a reader to successfully pursue experimental philosophy, McGuire and McMullin emphasize Newton's personal commitment to the simplicity, economy, and uniformity of nature to illuminate the connection between the general image of nature and natural bodies that is assumed in the *Principia* and the image of nature adopted by Newton's contemporaries. McGuire (1970) uses the commentary accompanying Rule 3 to situate Newton in the seventeenth-century debates about primary and secondary qualities. McMullin (1978) draws on that commentary to situate Newton in the early modern debates about the essences of bodies. The disadvantage of taking such an approach, as McGuire and McMullin

Using such an argument strategy limits the audience that Newton could persuade. For instance, a committed Cartesian, who prizes innate ideas and rational argument over the evidence-based inquiry that Newton endorses, would hardly be swayed by the remarks in the commentary to Rule 3. Nonetheless, the rhetorical strategy that Newton adopts allows him to bracket the questions about essences and real natures that must be addressed by those who opt to pursue a program of natural philosophy that promises truths and certainties. To do experimental philosophy, we don't need to have knowledge of what is really and actually the case in nature. Most notably, we don't need knowledge that whole bodies are really composed of insensible parts or that nature actually operates simply, economically, and uniformly. As characterized in the "Rules for the Study of Natural Philosophy," experimental philosophy is way of investigating nature that follows from these basic and common assumptions about the natural order of things. But it is not a way of investigating nature that depends on our ability to prove that these assumptions are true or indubitable. Indeed, according to Newton's remarks in the General Scholium, natural philosophy should not and cannot depend on our ability to know the real and actual order of nature, because, unlike God,

> we certainly do not know what is the substance of any thing. We see only the shapes and colors of bodies, we hear only their sounds, we touch only their external surfaces, we smell only their odors, and we taste their flavors. But there is no direct sense and there are no indirect reflected actions by which we know innermost substances....
>
> (*ibid*, 942)[65]

---

both acknowledge, is that the brief remarks from the commentary to Rule 3 do not provide a clear sense of why Newton finds atomism to be acceptable or of whether he considers extension hardness, impenetrability, mobility, and inertia to be primary and essential qualities of bodies. I avoid this specific interpretive difficulty, because, on my reading, Newton is not making any positive assertions about the real nature of bodies or the essential qualities of bodies. He is drawing attention to the reasoning that is standardly used when the qualities of extension, hardness, impenetrability, mobility, and inertia are universalized.

65 It would not be unreasonable to read the opening section of the General Scholium as Newton's defense of the image of nature and the evidence-based methodology that he associates with the practice of experimental philosophy. There, Newton connects God's nature with God's creation of the natural world, and he also contrasts God's abilities with our human abilities. What results are general characterizations of nature and of our human capacities that correspond to claims

For Newton, whatever empirical evidence we gather will always be limited. No matter what or how much evidence is disclosed to our senses, that evidence will simply not reveal what is or is not essential to natural bodies. Yet such limitations do not diminish what can be achieved in a natural philosophy that is founded on experimental and empirical evidence. In regard to gravity in particular, the best evidence available may never allow us to establish that it is essential to bodies; but as the argument that Newton forwards in Book 3 is meant to show, reasoning appropriately about empirical evidence can support taking gravity to be a quality that really exists and that belongs to all bodies universally.

Of course, whether in practice Newton has successfully distinguished his examination of empirical evidence from considerations of what is essential to bodies is a different matter. And in the case of gravity, it is a matter that hinges on Newton's success in bracketing questions about gravity's cause from questions about gravity's scope and characteristic features.

4.2 Gravity as Universal, Not Essential

You'll recall from Section 1.2 that in the Scholium to I.69 Newton claims that he is investigating attractive and impulsive forces found in nature in a very specific way. In the *Principia* he is not considering "the species of forces and their physical qualities," he tells us, but is instead focusing on "their quantities and mathematical proportions" (*ibid*, 588). Or, as he puts it in Definition 8, he is considering natural "forces not from a physical but only from a mathematical point of view" (*ibid*, 408). From this mathematical point of view, it does not matter whether an attractive force is produced by "spirits emitted" from the bodies, or by the "action of aether," or by "any medium whatsoever – whether corporeal or incorporeal – in any way impelling toward one another the bodies floating therein" (*ibid*, 588). As a consequence, in the *Principia* Newton uses the term "attraction" in "a general sense."

---

that Newton makes when presenting the Rules. However, the point I urge above remains intact, because, as in the commentaries accompanying the Rules, in the General Scholium Newton refrains from offering direct arguments for his basic premises. For instance, he offers no proof that God exists in the General Scholium. Instead, Newton notes that "It is agreed that the supreme God necessarily exists, and by the same necessity he is *always* and *everywhere*" (*ibid*, 942). He then spells out what follows from this agreed upon position.

It picks out "any endeavor whatever of bodies to approach one another" without designating what might be producing the force (*ibid*).[66]

Newton's argument for universal gravity exemplifies the noncausal approach to forces that he associates with the *Principia* as whole. As we saw in Sections 2.1 and 2.2, he establishes in Book 3 that gravity varies directly with mass and varies inversely as the square of the distance between two bodies, and as he demonstrates that gravity obeys these laws, he refrains from making any conjectures about the natural circumstances that might explain why such a force is found among natural bodies. In this respect, and fitting of the mathematical point of view that he adopts, Newton pursues his investigation into gravity without consideration of gravity's cause.

In the General Scholium, Newton offers an additional reason that he did not identify gravity's cause. He was pursuing experimental philosophy, not hypothetical philosophy, and consequently, his investigations into gravity could extend only as far as empirical and experimental evidence would allow. He thus does not isolate a particular cause for the force of gravity that he presents in Book 3, because there was insufficient evidence for him to do so. For instance, the evidence he had gathered revealed that "Gravity toward the sun is compounded of the gravities toward the individual particles of the sun, and at increasing distances from the sun decreases exactly as the squares of the distances as far out as the orbit of Saturn" (*ibid*, 943). But he has "not as yet been able to deduce from phenomena the reason for these properties of gravity" (*ibid*). In other words, the evidence he had available gave him no clear indication of what it is in nature that is producing a force with these features.

Newton does offer some general remarks about gravity's cause. He notes that whatever the cause of gravity is, it must be something that operates in such a way that it can produce the specific features of gravity that he has identified. "Indeed," he says,

> this force [of gravity] arises from some cause that [1] penetrates as far as the centers of the sun and the planets without any diminution of its power to act, and that [2] acts not in proportion to the quantity of the *surfaces* of the particles on which it acts (as mechanical causes are wont to do) but in proportion to the quantity of *solid* matter, and [3] whose action is extended everywhere to immense distances, [4] always decreasing as the squares of the distances.
>
> (*ibid*; bracketed terms and numbers added)

---

66 The relevant passages from the Scholium to I.69 and Definition 8 are quoted in full in Note 9.

The points that Newton makes here are relatively clear. He has determined that gravity is a quality that belongs to all bodies universally, including bodies that are immense distances from the earth, and so, whatever the cause of gravity is, it must be able to act "everywhere to immense distances." Additionally, having shown that the measure of gravity varies inversely as the square of the distance between bodies, where the distance is measured relative to the centers of these bodies, it follows that the cause of gravity must act accordingly. Namely, it must penetrate "as far as the centers of the sun and the planets without any diminution of its power to act," and it must "always decreas[e] as the squares of the distances" between the bodies on which it acts. He has also determined that the measure of gravity varies directly with the mass of bodies, and thus, the cause of gravity must act "in proportion to the quantity of *solid* matter," not in proportion to the surface area of "the particles on which it acts."[67]

In setting out the positive conditions that the cause of gravity must satisfy, Newton is at the same time ruling out the possibility that gravity could be explained by the causal model that Descartes presents in Part IV of the *Principles*. There Descartes claims that terrestrial bodies gravitate toward the earth (that is, they tend to fall toward the earth) because moving particles of ether are impacting their surfaces and pushing these bodies downward. Descartes additionally claims that the specific measure of a body's gravity does not depend on its quantity of matter; it depends only on how much the body's surface is impacted by the ether that surrounds it (Descartes 1984, 192; IV. 25; cf. Section 1.1). For Newton, this causal model simply won't do. He has demonstrated that the measure of a body's

---

67 In the closing paragraph of the General Scholium, Newton posits that "a certain very subtle spirit pervading gross bodies and lying hidden in them" might possibly explain why "the particles of bodies attract one another at very small distances and cohere when they become contiguous" (Newton 1999, 943–944). However, he does not provide a robust explanation for how this possible spirit might produce attraction or cohesion, and he doesn't because "there is not a sufficient number of experiments to determine and demonstrate accurately the laws governing the actions of this spirit" (*ibid*, 944). In Query 21 of the second English edition of the *Opticks* (1717/1718), Newton speculates that an aether of varying density might explain the presence of gravity in nature (Newton 1952, 350–353). But, as in the General Scholium, he neither asserts nor argues for the existence of this possible physical cause. Instead, and in line with his explicit claim in Query 21 that "I do not know what this Æther is," Newton's remarks about the cause of gravity in the *Opticks* are most naturally read as a conditional statement. He is showing that *if* one were to try to explain gravity in terms of an aether, *then* the aether would have to have a particular set of features.

gravity is directly proportional to its mass and does not vary according to the body's surface area. This means that, whatever the cause of
gravity is, it cannot behave as Descartes's ether. Or, as Newton puts
it, whatever it is that causes the measure of a body's gravity to be proportional to the body's mass cannot act "in proportion to the quantity
of the *surfaces* of the particles on which it acts (as mechanical causes
are wont to do)"; it must instead act "in proportion to the quantity of
solid matter" (*ibid*).

In drawing this conclusion, Newton has not only ruled out
Descartes's explanation for gravity; it seems he has also ruled out
the general possibility that gravity could be produced by any sort of
mechanical cause. For readers of the *Principia* who came to the text
committed to a mechanical view of natural bodies, such a suggestion
was sufficient reason to reject Newton's notion of gravity. Though
few of them denied that God could change nature's normal course by
supernatural intervention, for these readers, the motions and tendencies to motions commonly found among natural bodies could only be
produced through mechanical means. More precisely, they worked
under the assumption that all natural motions and all natural tendencies to motion had to be explainable in terms of contact action, that
is, in terms of the impacts suffered by the surface of a body. It was on
these grounds that commentators such as Huygens and Leibniz ob-
jected to the very idea that Newton's gravity *really exists* in nature and
could serve as the basis for natural philosophy.[68] From their standpoint, there would be no harm in accepting Newton's notion of gravity
as a hypothetical force, as a posit that allows us to make successful
predictions and thereby save the phenomena. However, if we accept
Newton's gravity is a real quality of bodies, and one that cannot be
explained in terms of nature's mechanical operations, then, according to Leibniz, we must also accept that it is "an unreasonable" and
"a very occult" quality of bodies. In other words, if Newton's gravity
really exists in nature, it must exist as "a simple primitive Quality"
that is essential to and inherent in a body, and that itself produces
motions "without any intelligible Means" (Leibniz 1712, 139).[69]

---

68 See Note 12 for texts in which Huygens and Leibniz forward this criticism.
69 According to Leibniz, a "reasonable" explanation just is a mechanical explanation
    such that if an explanation is not cast in terms of impacts and contact action, it is
    not a sufficient reason for a natural event. Leibniz urges this point in the portion
    of his May 1712 letter to Hartsoeker from which I quote above. The allegation
    that the gravity of the *Principia* ought to be considered an inherent and essential
    quality of bodies was not unique to Newton's critics. The position was also voiced

In response to this line of criticism, Newton added four sentences to the end of the commentary accompanying Rule 3. These sentences appear only in the third edition *Principia* and read as follows:

> Yet I am by no means affirming that gravity is essential to bodies. By inherent force I mean only the force of inertia. This is immutable. Gravity [*Gravitas*] is diminished as bodies recede from the earth.
> *(ibid,* 796; bracketed Latin term added)[70]

Put in the most positive light, Newton is not (or not merely) insisting that his critics have attributed to him a position that he did not explicitly endorse. He is insisting that, contrary to the suggestion of his critics, the presentation of gravity in the *Principia* does not demand that we think of it as a force that is inherent in and essential to bodies. Accordingly, the question of whether Newton's reply is *effective* is a question of whether he has presented sufficient evidence in the *Principia* to undermine the claim that gravity's real existence as a universal quality of bodies can only be explained if it is understood as an essential and inherent quality of bodies. And from this standpoint, both the contrast that Newton draws between inertia and gravity and the meaning of Rule 3 take on special significance.[71]

---

by more sympathetic commentators who accepted Newton's claim that "gravity really exists." For instance, in the "Editor's Preface to the Second Edition," Roger Cotes describes the gravity of Book 3 as a "primary quality" of bodies and says that "Among the primary qualities of all bodies universally, either gravity will have a place, or extension, mobility, and impenetrability will not" (Newton 1999, 392). Immanuel Kant makes a stronger point in his *Metaphysical Foundations of Natural Philosophy* (1786) and maintains that Newton's universal gravity is a force of attraction that is "essential to all matter" (Kant 2004, 54). He also claims that Newton's denial that gravity is an "original" and essential attraction "set [Newton] at variance with himself" (*ibid,* 54). For discussion of Kant's criticism of Newton, see Friedman (1990). For discussion of the approach Kant takes to universal gravity in the *Metaphysical Foundations,* see Friedman (2013), especially Section 2.18.

70 "Attamen gravitatem corporibus essentialem esse minime affirmo. Per vim insitam intelligo solam vim inertæ. Hæc immutabilis est. Gravitas recedendo a terra diminuitur" (Newton 1871, 389).

71 Newton also explicitly denies that the force of gravity presented in the *Principia* should be considered essential to bodies in two letters that he wrote to Richard Bentley in early 1692/3. In the letter from 17 January 1692/3, Newton says to Bentley,

> You sometimes speak of gravity as essential and inherent to matter. Pray do not ascribe that notion to me; for the cause of gravity is what I do not pretend to know, and therefore would take more time to consider of it.
> (Newton 2004, 100).

Consider first Newton's remark that "By inherent force I mean only the force of inertia. This is immutable." There's no special difficulty mapping these claims on to what's presented in the *Principia*. In all editions of the text, inertia is defined as an "inherent force of matter," and it is also characterized as immutable. According to Definition 3:

> *Inherent force of matter is the power of resisting by which every body, so far as it is able, perseveres in its state either of resting or of moving uniformly straight forward.* **This force is always proportional to the body and does not differ in any way from the inertia of the mass except in the manner in which it is conceived.** Because of the inertia of matter, every body is only with difficulty put out of its state either of resting or of moving. Consequently, **inherent force may also be called by the very significant name of force of inertia.**
>
> (*ibid*, 404; boldface added)

Now, according to Definition 1, a particular quantity of matter picks out a particular body.[72] And according to Definition 3, where there is a particular body, or a particular quantity of matter, there is inertia. In-

---

In the letter from 25 February 1692/3, Newton puts the point more strongly:

> That gravity should be innate, inherent, and essential to matter, so that one body may act upon another at a distance through a vacuum without the mediation of anything else, by and through which their action and force may be conveyed from one to another, is so great an absurdity, that I believe no man who has in philosophical matters a competent faculty of thinking can ever fall into it. Gravity must be caused by an agent acting constantly according to certain laws; but whether this agent be material or immaterial, I have left to the consideration of my readers.
>
> (Newton 2004, 102–103)

From these remarks to Bentley, it is relatively clear that Newton denies that *bodies* can act on each other at a distance. However, following John Henry (2004, 2011), these remarks leave open the question of whether Newton also denies that *the cause of gravity* can act at a distance. According to Henry, Newton makes no such denial. Commentators such as Janiak (2008) and Kochiras (2009) claim that he does. On the question of Newton's attitude toward action at a distance, see also Ducheyne (2011) and Schliesser (2011).

72 "Definition 1: *Quantity of matter is a measure of matter that arises from its density and volume jointly...*For the present, I am not taking into account any medium, if there should be any, freely pervading the interstices between the parts of bodies. Furthermore, I mean this quantity whenever I use the term "body" or "mass" in the following pages" (Newton 1999, 403–404). See Brading (2012) for discussion of how the three laws of motion contribute to the conception of body that Newton adopts in the *Principia*.

ertia is thus an inherent force insofar as it is an enduring, irremovable feature of every body, that is, of every quantity of matter, and in this respect, it is essential to bodies. Newton's additional claim that inertia "is immutable" also follows from these definitions. According to Definition 1, a particular body just is a particular quantity of matter. And according to Definition 3, inertia is "always proportional to the body," that is, it always varies directly with mass, or quantity of matter. Consequently, the specific measure of inertia for one and the same body – for one and the same particular quantity of matter – cannot vary in degree, and therefore, in this respect, inertia "is immutable."

In the *Principia*, gravity is presented differently. It is not by definition that gravity is proportional to a body's mass and inversely proportional to the square of the distance between two bodies. And it is not by definition that gravity is taken to be a universal quality of bodies. According to its most basic definition, gravity is the tendency that a body has to fall toward some other body. Identifying gravity's additional, more specific features and demonstrating that it belongs to all bodies universally requires the arguments of Book 3 – arguments, as we've seen, that involve the use of experimental and empirical evidence, inductive generalizations, and the four Rules for the Study of Natural Philosophy. On these grounds alone, it seems we have reason to resist considering gravity an inherent and essential quality of bodies. With the laws and universality of gravity presented as propositions that have been "gathered from phenomena by induction," as per Rule 4, there lingers the possibility that "yet other phenomena" could challenge what has been demonstrated of gravity, including its status as a universal quality of bodies (*ibid*, 796). No such possibility lingers with inertia, at least not if Definitions 1 and 3 stipulate how "body" and "inertia" must be understood in order to pursue the natural philosophy of the *Principia*. If those definitions are taken to be constitutive of Newton's program of experimental philosophy, then inertia will remain an inherent and essential force of all bodies, empirical evidence come what may.

That the difference between how inertia and gravity are presented in the *Principia* indicates that gravity should not be considered inherent or essential is implied by Newton's remarks. What he says *explicitly* is that inertia is immutable, whereas "Gravity is diminished as bodies recede from the earth" (*ibid*, 796). The general point Newton is making in response to his critics seems relatively straightforward. He is pointing out that the gravity he has presented in the *Principia* is a force with a measure that varies in degree and that varies, specifically, in relation to the distance between bodies. He is thus pointing out that the gravity he has presented in the *Principia* is not immutable in the sense that inertia is. The gravity of one and the same body does vary in degree; the inertia of one and the same body does not. And what Newton is suggesting to his Leibnizian readers is that because the specific measures of a body's gravity vary, we should not take gravity to be an inherent or essential force of bodies.

Now, to get from a quality that varies in degree to a quality that should not be considered inherent and essential seems to require a bit of leap on Newton's part. However, the leap is not unjustified, given his remarks from the beginning of the commentary to Rule 3. You'll recall that in the one-sentence justification for the Rule that opens the commentary, Newton reports that if empirical evidence shows us that a quality of bodies "cannot be diminished," then the quality can be understood as one that "cannot be taken from bodies" (*ibid*, 795). To emphasize at the very end of the commentary that the specific measures of a body's gravity can be diminished is thus to point out that there is inadequate evidence for considering it a quality of bodies that cannot be taken away, and thus, that there is inadequate evidence for considering it to be the sort of quality that could count as inherent in and essential to bodies. By also offering the remark that inertia is immutable, Newton is additionally suggesting that, of the two forces, there is only adequate evidence for considering inertia as inherent and essential. Given how this force is defined, this suggestion makes sense. As we just saw, the specific measure of a body's inertia remains invariant, which means that inertia cannot be diminished. This means in turn that inertia *can* be regarded as a quality that cannot be taken away from bodies, and therefore as a quality that is inherent in and essential to bodies. Newton's critics might balk at the claim that qualities that cannot be diminished should be understood as qualities that cannot be taken away from bodies, but by Newton's standards at least, there is merit to his claim that the variation that characterizes the specific measures of a body's gravity gives us reason to resist considering gravity to be an inherent and essential force.

However, depending on how we read Rule 3, there is a question of whether, with his reply, Newton has also given us reason to not consider gravity as a *universal* quality of bodies. If his central claim is that empirical evidence does not support us thinking of gravity as a quality that cannot be taken away from bodies, then, in accordance with the one-sentence justification for Rule 3, it follows that empirical evidence also does not support us regarding gravity as a quality of bodies that cannot be intended and remitted. On the One-Set Reading of Rule 3, it would additionally follow that gravity should not be taken as a quality of all bodies universally, because what Newton has asserted is that gravity fails to meet one of the two sufficient and necessary conditions for universalizing a quality that the Rule identifies.

We saw in Section 2.3 that commentators who adopt the One-Set Reading have a way to address this problem. The key for them is to expand the meaning of "qualities of bodies that cannot be intended and remitted" so that it picks out qualities that are invariant proportions and also qualities with general measures characterized by invariant proportions. Taking this tack, the variation in the *specific* measures of a body's gravity does not undermine gravity's status as a universal quality. Since evidence indicates that gravity's *general* measure is characterized by invariant proportions – that its general measure always varies directly with mass and always varies inversely as the square of the distance between two bodies – evidence indicates that gravity should be regarded as a quality that cannot be intended and remitted.

But now an additional question lingers. If it is the invariance that characterizes gravity's general measure that allows us to count it as a quality of bodies that cannot be intended and remitted, why is this not also adequate reason for us to count gravity as a quality that cannot be taken away from bodies and, in turn, to regard it as inherent in and essential to bodies? It seemed that the point of Newton emphasizing that inertia is immutable was to highlight that the very feature that allows us to consider inertia as a quality that cannot be intended and remitted is also the feature that allows us to consider it inherent in and essential to bodies. So, if there is evidence that gravity cannot be intended and remitted, isn't there also evidence that it cannot be diminished and, thus, evidence that it cannot be taken away from bodies? And if so, wouldn't this same evidence indicate that gravity, like inertia, can be considered an inherent and essential quality of bodies?

Granted, on the One-Set Reading, the evidence that inertia and gravity cannot be intended and remitted is different in kind. In the case of inertia, the specific measure of inertia for one and the same body is what cannot increase or decrease. In the case of gravity, the mathematical relations that gravity bears to mass and to distance are what cannot be diminished. Moreover, of the two forces only gravity bears a relation to distance, and it is this relation that explains the variation in gravity's specific measures. But these differences leave intact the central puzzle here. Since there is some detectable invariance that characterizes both inertia and gravity – and, indeed, both have an invariance that stems from their direct proportionality to mass – and since, on the One-Set Reading, it is this invariance that indicates that each is a quality that cannot be intended and remitted, it is not clear why Newton would claim that *only* inertia's invariance gives us grounds for regarding it as a quality that is inherent in and essential to bodies. With this question lingering, Newton's reply to his critics wouldn't necessarily be ineffective. But without some additional explanation, at minimum, his reply would remain incomplete.[73]

On the Two-Set Reading of Rule 3, no such questions linger, because, on this reading, Newton does not claim that gravity is a quality that cannot be intended and remitted. Instead, Newton maintains the medieval usage of *intendi & remitti*, and when he notes at the end of the Rule's commentary that gravity "is diminished as bodies recede from the earth," he is noting that it is a quality that can be intended and remitted. In line with the remark that opens the commentary, the more specific point that he is making in reply to his critics is that insofar as gravity can be diminished, there is no evidentiary basis for claiming that gravity cannot be taken away from bodies, and thus that there is no evidentiary basis for considering gravity a quality that is inherent in and essential to bodies. And by making this claim, Newton does not thereby motivate questions about gravity's status as universal. On the Two-Set Reading, gravity remains a quality that can be taken as a quality of all bodies universally, because it meets the other condition for universalization that Rule 3 identifies. Namely, given the store of empirical evidence that is used to complete the argument for universal gravity in Book 3, there are adequate evidentiary grounds for regarding it as a quality that can be found among all bodies on which experiments can be made.

If this is the appropriate way to read Newton's reply to his critics – if his central point is that gravity should not be considered an inherent and essential quality of bodies because there is inadequate evidence that it cannot be taken away from bodies (and thus inadequate evidence that it cannot be intended and remitted) – we have a way of understanding why Newton would cast Rule 3 as a claim expressing two sufficient but not necessary conditions. By specifying in the Rule two different but possibly overlapping sets of universalizable qualities,

---

73 I have not come across any commentaries that put the problem I raise here in precisely the same terms that I use. However, an analogue can be found in Okruhlik (1989), who casts the problem in causal terms (cf. Okruhlik 1989, 111–112). She suggests that even if the critics grant Newton the distinctions that he draws between inertia and gravity, they would still maintain that there could be no mechanical explanation of the invariant relationship that gravity bears to distance, and thus no way to explain gravity's presence in nature except by considering it an inherent and essential feature of bodies. In the terms I use, the critics would be demanding an explanation of why we ought to accept that the invariant relationship that gravity bears to distance does not indicate that it is inherent and essential.

he has given the experimental philosopher a directive that allows her to draw empirically well-supported inferences about which qualities belong to all bodies universally and to do so without also having to conclude that those qualities should be regarded as inherent in and essential to bodies. For, on the reading above, what Newton's reply to his critics indicates is that only qualities like inertia – only those that meet *both* of the conditions set out by the Rule – ought to be taken as universal and also as inherent and essential.[74] Qualities like gravity should be understood differently. Insofar as there is evidence that indicates that they are general qualities, they can be regarded as universal qualities. But insofar as there is no evidence that they are qualities that cannot be taken away from bodies, they should not also be taken to be inherent and essential qualities of bodies. In this respect, on the Two-Set Reading, Newton has with Rule 3 set a high evidentiary bar for considering qualities of bodies to be universal and set an even higher evidentiary bar for considering them to be universal qualities that are inherent in and essential to bodies.

4.3 Qualities and Explanations in Experimental Philosophy

You'll notice that on the reading I offer above, Newton is not alleging to have knowledge of how nature really and actually operates as he responds to his critics. He is not claiming that the world is such that inertia *is* an inherent and essential quality of bodies and gravity is *not*. He is emphasizing that there is inadequate evidence to support the critic's claim that, based on what has been demonstrated in the *Principia*, we must infer that gravity is an inherent and essential quality of bodies.[75]

---

74 That inertia fulfills both of the conditions specified by Rule 3 follows from Definition 3. On the one hand, according to that definition inertia is "always proportional to the body." As noted above, this means that the specific measure of inertia for one and the same body cannot be diminished, which, by Newton's standards, means in turn that inertia is a quality that cannot be taken away from bodies and, thus, is a quality that cannot be intended and remitted. On the other hand, according to Definition 3 inertia is "an inherent force of matter," and since bodies just are quantities of matter, inertia will be a quality of all bodies, including all those bodies on which experiments can be made.

75 In all editions of the *Principia*, inertia is defined as an inherent force of matter, but to my knowledge, there is no text in which Newton affirms that inertia is in fact essential to bodies. In the pre-1713 draft version of Rule 3 that was mentioned earlier, Newton does include inertia among those qualities "which cannot be intended and remitted" and that "cannot be taken away from bodies" (cited in McGuire 1968,

This same general theme runs throughout the "Rules for the Study of Natural Philosophy." In presenting Rules 1, 2, and 3, Newton focuses on what inferences the experimental philosopher should draw when attempting to identify the causes of natural effects and the qualities that belong to all bodies. In presenting Rule 4, he focuses on what should be inferred about propositions that have been "gathered from phenomena by induction" and claims they "should be considered either exactly or very nearly true notwithstanding any hypotheses" (Newton 1999, 796). These directives stem from a commitment to more basic assumptions – assumptions concerning nature's simple, economical, and uniform order; the priority of experimental and empirical evidence in our investigations of nature; and the generally appropriate ways to reason about that evidence. But neither these premises nor the directives that follow from them are forwarded by Newton as claims that are true or that can serve as the means for discovering what is true about the actual order of nature.

Newton's resistance to making claims about what is and is not true of nature is connected with the rejection of hypotheses that is part and parcel of his experimental philosophy. As he sets forth the appropriate ways to practice natural philosophy, Newton's fundamental message is that our positive claims about nature must be based on what empirical evidence reveals and that our reasoning about this evidence must remain faithful to nature's simple, economical, and uniform ordering. And insofar as our senses are not equipped to give us insight into the essences or innermost substances of things, to speculate about what is essential to bodies, or, more generally, to speculate about what mechanisms are producing the empirical evidence that is available, is to venture into a territory where the limits of our sensory capacities and the guidance of empirical evidence are ignored. It is to rely on "idle fancies," or "dreams and vain fictions of our own devising," and

---

237, with the Latin text provided on 257; cf. Note 55 for the full quotation). McGuire interprets Newton as additionally claiming in this draft that qualities that cannot be taken away from bodies are essential qualities, which would be consistent with the other remarks from that same period in which Newton identifies the empirical evidence that allows the experimental philosopher to make positive claims about the actual order of nature (cf. Note 21). However, McGuire acknowledges that it is difficult to draw firm conclusions about Newton's stance toward essential qualities from the draft, because the manuscript is heavily damaged. Also, in the parts that are legible, there are several places where Newton adds and cancels text, and it is not always clear what exactly Newton wanted to retain and what he wanted to omit (cf. McGuire 1968, 236).

for Newton, relying on such hypotheses is simply unacceptable when practicing experimental philosophy.

The rejection of hypotheses is one key difference between the program of natural philosophy that Newton pursues in the *Principia* and the hypothetical philosophy that Descartes pursues in the *Principles*. We have seen other important points of contrast as well. Descartes attributes gravity only to terrestrial bodies. Newton argues that gravity belongs to all bodies universally. Descartes refers to the motions of a material ether to explain terrestrial gravity and to explain the observed motions of planetary bodies. Newton does not posit a material ether to explain any observed motions, whether of terrestrial bodies or of celestial bodies. There are also differences in how Descartes and Newton defend their respective ways of practicing of natural philosophy. Descartes argues for the superiority of his hypothetical philosophy by showing his readers that it is based on clear and evident first principles from which additional clear and evident truths about nature can be deduced. Newton argues for the superiority of his experimental philosophy by showing his readers that it is premised on an image of nature and an attitude toward empirical evidence that he assumes his readers already accept as basic and uncontroversial.

As we have also seen, Descartes and Newton hold different conceptions of explanation. For Descartes, to explain the motions and tendencies to motion that we observe requires some account, even if hypothetical, of the mechanism that is producing these motions and tendencies. For Newton, there is no such requirement. In experimental philosophy, we are to isolate the questions that are to be answered based on empirical evidence from questions about the reasons for, or causes of, this evidence. Newton's argument for universal gravity exemplifies this sort of noncausal approach, and given the ground we have covered, that argument also sheds light on the different ways in which Newton and Descartes isolate the qualities of bodies that can be used to explain natural philosophy. To bring this difference into fuller relief, we have to turn back to Descartes's *Principles* one last time.

You'll recall that, for Descartes, all explanations of the visible world must be consistent with the first principles of his natural philosophy, including the principle of corporeal things, namely, "that there are bodies extended in length, width, and depth, which have diverse figures and are moved in diverse ways" (Descartes 1984, xxii). The explanations that Descartes forwards in Part III of the *Principles* fulfill this general requirement: They are explanations premised on extension being the essential quality of natural bodies. But more specifically, these explanations are cast in terms of qualities that derive from the quality of extension, namely, they are cast in terms of the divisions, shapes, and motions of natural bodies.[76] As Descartes explains in the final section of Part II,

> I openly acknowledge that I know of no kind of material substance other than that which can be divided, shaped, and moved in every possible way, and which Geometers call quantity and take as the object of their demonstrations. And {I also acknowledge} that there is absolutely nothing to investigate about this substance except those divisions, shapes, and movements.
>
> (*ibid*, 77; II.64)

The central role of motion in Descartes's explanations of observed phenomena is clear from his Vortex Hypothesis. According to this model, the observed motion of the planets around the sun is explained in reference to the swirling motion of a material ether that carries along these heavenly bodies. The central role that Descartes assigns to the divisions and shapes of natural bodies is more evident in other cases. For instance, he explains the observed behavior of fire by imagining fire's particles to have a particular shape and motion. He also posits that there is a specific way that these particles can be divided and also a specific way that these particles can divide other bodies (cf. *ibid*, 121–126; III.70–77).[77]

Generally speaking, there is nothing unreasonable about Descartes's decision to use the divisions, shapes, and motions of natural bodies to explain observed natural events. They are nonessential qualities of natural bodies, but as qualities that derive from the essential quality of extension, they are nonessential qualities that we can be assured are common to all natural bodies. And it is on this specific point – on the question of what sort of evidence is necessary for the natural philosopher to be assured that a quality is common to natural

---

76 According to Descartes's technical terminology, extension would be considered an *attribute* of material bodies, because extension belongs to the nature of bodies. Divisions, shapes, and motions would instead be considered *modes* of material bodies, because these can be altered without also altering the body's status as a fundamentally extended thing. See Descartes (1984, 24–25; I.55–57). I use the term "quality" in my presentation to illuminate how Descartes's position can be connected with Newton's, and I use that term in a generic sense to refer to any feature that belongs to a material body.

77 See McMullin (2008, 2009) and Domski (2019) for discussion of the role played by imaginative constructs in the explanations that Descartes forwards in Part III of the *Principles*.

bodies – that we can now better appreciate the broader significance of Rule 3 for the methodology that Newton associates with the practice of experimental philosophy.

For Newton, of course, any evidence that is used in experimental philosophy must be empirical. Nowhere in the *Principia* does he entertain the possibility that we could follow Descartes and solely use what is rationally clear and evident to gain knowledge of existing things. Newton also parts ways with Descartes on the question of whether it is necessary to know the essences of natural bodies to explain natural phenomena. Still, Newton shares Descartes's general position that our natural philosophical explanations of phenomena are to be cast in terms of qualities that we can be assured are common to natural bodies. On the reading I've offered above, he also shares the related position that these common qualities need not be essential qualities. Descartes isolates divisions, shapes, and motions as the nonessential but common qualities to be used to explain the phenomena of the visible world. Newton instead uses a universal but nonessential force of gravity.

We saw in Sections 2.1 and 2.2 that the Rules for the Study of Natural Philosophy play a key role in Newton's argument for universal gravity and allow him to reason from bodies on which experiments *have been made* to bodies on which experiments *can be made* and, ultimately, to celestial bodies on which experiments *cannot be made*. In that argument, Rule 3 is applied just once, in Corollary 2 of III.6. But the general progression of the argument exemplifies what the Rule communicates, namely, that under appropriate evidentiary circumstances, a quality of bodies should be taken to be a quality of all bodies universally. As noted in Section 1.2, it is because what Newton does in Book 3 maps on to what Rule 3 says that scholars have come to view the Rule as "a primary statement of Newton's philosophy of science" (Cohen 1971, 24). We can now also view the Rule as expressing a primary difference between Newton's and Descartes's approaches to the universal qualities of natural bodies. Whereas for Descartes universal qualities are to be "deduced" from the essential quality of extension, with Rule 3 Newton instructs us instead to use empirical evidence to identify the qualities that are common to all natural bodies, and notably, the Rule allows the experimental philosopher to universalize qualities even when the available evidence does not reveal why these qualities are common to all bodies.

Put in this light, Rule 3 helps clarify how Newton was able to take a further step away from Descartes's natural philosophy and widen the scope of the qualities that can play an explanatory role in natural philosophy. For Newton, a quality used to explain some set of phenomena must be common to those phenomena, but, on the Two-Set Reading of Rule 3, Newton makes no Cartesian demand that such a quality must also be taken to be essential to bodies or to derive from an essential quality. On the Two-Set Reading, Rule 3 tells us that two different but possibly overlapping sets of qualities can be universalized. There are those that experimental and empirical evidence indicates are qualities that cannot be intended and remitted and those that experimental and empirical evidence indicates are qualities found among all bodies on which experiments can be performed. The evidence need not show why a quality belongs to these sets, and as signaled by the distinction Newton draws between inertia and gravity, only when there is evidence that a quality belongs to both sets are there grounds for claiming that the quality is inherent in and essential to bodies. But, contra Descartes, evidence that a quality is inherent in and essential to bodies is not necessary to take the quality to be a quality of all bodies universally.

Ultimately, on the reading I've offered above, this is precisely why a nonessential force of gravity could serve as the central element in Newton's explanations of celestial and terrestrial motions. Newton produced experimental, evidence-based arguments to demonstrate that gravity should be regarded as a quality common to all natural bodies – or, in his terms, that it should be accepted as a force that "exists in all bodies universally." The evidence that he used did not indicate why gravity is common – it did not disclose why gravity "square[s] with experiments generally" or, more generally, how gravity could belong to all natural bodies. But in experimental philosophy, this makes no matter. In experimental philosophy, successfully explaining the phenomena by means of natural forces does not require identifying the causes of these forces. It requires reasoning from experimental and empirical evidence and producing arguments that extend only so far as that evidence allows. This is why no cause of gravity is identified, and this is also why an experimental philosopher will remain satisfied with the explanations of observed motions that are presented in Book 3. For this non-Cartesian sort of natural philosopher – who grounds her investigations into nature on experimental and empirical evidence; who reasons on analogy with nature's simple, economical, and uniform ordering; and who does not use hypotheses to explain the phenomena – "it is enough that gravity really exists and acts according to the laws that [Newton has] set forth and is sufficient to explain all the motions of the heavenly bodies and of our sea" (*ibid*, 943; bracketed phrase added).

References

Belkind, Ori (2017). "On Newtonian Induction." *Philosophy of Science* 84 (4): 677–697.

Bertoloni Meli, Domenico (2006). *Thinking with Objects: The Transformation of Mechanics in the Seventeenth Century.* Baltimore, MD: The Johns Hopkins University Press.

Biener, Zvi (forthcoming). "Newton's Regulae Philosophandi." In Eric Schliesser and Christopher Smeenk, eds., *Oxford Handbook of Isaac Newton.* New York and Oxford: Oxford University Press. Retrieved from https://zbiener.github.io/files/Biener2018a.pdf (Accessed: 14 October 2019).

Biener, Zvi and Mary Domski (forthcoming). "Working Hypotheses and the Logic of Theory-Mediation in the *Principia.*" In Eric Schliesser, Christopher Smeenk, and Marius Stan, eds., *Theory, Evidence, Data: The Philosophy of George E. Smith* (Boston Studies in the Philosophy and History of Science). Dordrecht: Springer.

Biener, Zvi and Eric Schliesser, eds. (2014). *Newton and Empiricism.* New York and Oxford: Oxford University Press.

Biener, Zvi and Eric Schliesser (2017). "The Certainty, Modality, and Grounding of Newton's Laws." *Monist* 100 (3): 311–325. To be reprinted in Eric Schliesser (forthcoming), *Newton's Metaphysics*, New York and Oxford: Oxford University Press.

Biener, Zvi and Christopher Smeenk (2012). "Cotes' Queries: Newton's Empiricism and Conceptions of Matter." In Andrew Janiak and Eric Schliesser, eds., *Interpreting Newton: Critical Essays.* Cambridge, UK and New York: Cambridge University Press, pp. 103–137.

Brading, Katherine (2012). "Newton's Law-Constitutive Approach to Bodies: A Response to Descartes." In Andrew Janiak and Eric Schliesser, eds., *Interpreting Newton: Critical Essays.* Cambridge, UK and New York: Cambridge University Press, pp. 13–32.

Cohen, I. Bernard (1966). "Hypotheses in Newton's Philosophy." *Physis. Rivista Internazionale di Storiadella Scienza* 8: 163–184.

Cohen, I. Bernard (1971). *Introduction to Newton's 'Principia.'* Cambridge, UK: Cambridge University Press.

Cohen, I. Bernard (1999). *A Guide to Newton's* Principia. In Newton (1999), pp. 1–370.

Descartes, René (1984). *Principles of Philosophy*. Translated, with explanatory notes by V. R. Miller and R. P. Miller. Dordrecht: D. Reidel Publishing Company.

Domski, Mary (2003). "The Constructible and the Intelligible in Newton's Philosophy of Geometry." *Philosophy of Science* 70 (5): 1114–1124.

Domski, Mary (2018). "Laws of Nature and the Divine Order of Things: Descartes and Newton on Truth in Natural Philosophy." In Walter Ott and Lydia Patton, eds., *Laws of Nature*. New York and Oxford: Oxford University Press, pp. 42–61.

Domski, Mary (2019). "Imagination, Metaphysics, Mathematics: Descartes's Arguments for the Vortex Hypothesis." *Synthese* 196 (9): 3505–3526.

Ducheyne, Steffen (2011). "Newton on Action at a Distance and the Cause of Gravity." *Studies in History and Philosophy of Science, Part A* 42 (1): 154–159.

Ducheyne, Steffen (2012). *The Main Business of Natural Philosophy: Isaac Newton's Natural-Philosophical Methodology*. Dordrecht: Springer.

Friedman, Michael (1990). "Kant and Newton: Why Gravity is Essential to Matter." In Phillip Bricker and R. I. G. Hughes, eds., *Philosophical Perspectives on Newtonian Science*. Cambridge, MA and London, England: The MIT Press, pp. 185–202.

Friedman, Michael (2013). *Kant's Construction of Nature: A Reading of the Metaphysical Foundations of Natural Science*. Cambridge, UK and New York: Cambridge University Press.

Guicciardini, Niccolò (2009). *Isaac Newton on Mathematical Certainty and Method*. Cambridge, MA and London, England: The MIT Press.

Henry, John (2004). "'Pray Do Not Ascribe That Notion to Me': God and Newton's Gravity." In J. E. Force and R. H. Popkin, eds., *The Books of Nature and Scripture: Recent Essays on Natural Philosophy, Theology and Biblical Criticism in the Netherlands of Spinoza's Time and the British Isles of Newton's Time*. Dordrecht: Kluwer Academic Publishers, pp. 123–147.

Henry, John (2011). "Gravity and *De Gravitatione*: The Development of Newton's Ideas on Action at a Distance." *Studies in History and Philosophy of Science, Part A* 42 (1): 11–27.

Iliffe, Rob and George E. Smith, eds. (2016). *The Cambridge Companion to Newton*. Second Edition. Cambridge, UK: Cambridge University Press.

Janiak, Andrew (2008). *Newton as Philosopher*. New York: Cambridge University Press.

Janiak, Andrew and Eric Schliesser, eds. (2012). *Interpreting Newton: Critical Essays*. Cambridge, UK and New York: Cambridge University Press.

Kant, Immanuel (2004). *Metaphysical Foundations of Natural Science*. Translated and edited by Michael Friedman. Cambridge, UK: Cambridge University Press.

Kochiras, Hylarie (2009). "Gravity and Newton's Substance Counting Problem." *Studies in History and Philosophy of Science, Part A* 40 (3): 267–280.

Koyré, Alexandre (1965). *Newtonian Studies.* Chicago, IL: The University of Chicago Press.

Leibniz, G. W. (1712). "Philosophical Letters Written by M. Leibniz and M. Hartsoeker." *Memoirs of Literature* 2 (18): 137–143.

Leibniz, G. W. (1989). *Philosophical Essays.* Edited and translated by Roger Ariew and Daniel Garber. Indianapolis, IN: Hackett Publishing Company.

Mandelbaum, Maurice (1964). *Philosophy, Science, and Sense Perception: Historical and Critical Studies.* Baltimore, MD: The Johns Hopkins Press.

McGuire, J. E. (1968). "The Origin of Newton's Doctrine of Essential Qualities." *Centaurus* 12 (4): 233–260.

McGuire, J. E. (1970). "Atoms and the 'Analogy of Nature': Newton's Third Rule of Philosophizing." *Studies in History and Philosophy of Science, Part A* 1 (1): 3–58.

McMullin, Ernan (1978). *Newton on Matter and Activity.* Notre Dame, IN and London: University of Notre Dame Press.

McMullin, Ernan (2008). "Explanation as Confirmation in Descartes's Natural Philosophy." In Janet Broughton & John Carriero, eds., *A Companion to Descartes.* Malden, MA and Oxford: Blackwell Publishing, pp. 84–102.

McMullin, Ernan (2009). "Hypothesis in Early Modern Science." In Michael Heidelberger & Gregor Schiemann, eds., *The Significance of the Hypothetical in the Natural Sciences.* Berlin: Walter de Gruyter, pp. 7–37.

Newton, Isaac (1871). *Isaac Newton's* Principia. Facsimile of the Latin third edition (1726) text, reprinted for William Thomson and Hugh Blackburn. Glasgow, Scotland: James Maclehose, Glasgow, Publisher to the University. Retrieved from Online Library of Liberty at https://oll.libertyfund.org/titles/newton-principia-mathematica-latin-ed (Accessed: 14 October 2019).

Newton, Isaac (1934). *Sir Isaac Newton's* Mathematical Principles of Natural Philosophy and His System of the World. Two Volumes. Translated into English by Andrew Motte in 1729. The translations revised, and supplied with an historical and explanatory appendix, by Florian Cajori. Berkeley, Los Angeles, and London: University of California Press.

Newton, Isaac (1952). *Opticks, or A Treatise of the Reflections, Refractions, Inflections and Colors of Light.* Based on the fourth edition of 1730. New York: Dover Publications, Inc.

Newton, Isaac (1999). *The* Principia: *Mathematical Principles of Natural Philosophy.* Based on the third edition of 1726. Translated by I. Bernard Cohen and Anne Whitman. Berkeley, Los Angeles, and London: University of California Press.

Newton, Isaac (2004). *Isaac Newton: Philosophical Writings.* Edited by Andrew Janiak. Cambridge, UK: Cambridge University Press.

Newton, Isaac (2009). *Philosophiae Naturalis Principia Mathematica.* The Latin first edition of 1687. Released as an open-access E-Book by The Project Gutenberg. Produced by Jonathan Ingram, Keith Edkins and the Online Distributed Proofreading Team at http://www.pgdp.net. Retrieved from The Project Gutenberg at http://www.gutenberg.org/files/28233/28233-pdf.pdf (Accessed: 14 October 2019).

Okruhlik, Kathleen (1989). "The Foundation of All Philosophy: Newton's Third Rule." In James Robert Brown and Jürgen Mittelstrass, eds., *An Intimate Relation*. Boston Studies in the Philosophy of Science, Vol. 116. Dordrecht: Kluwer Academic Publishers, pp. 97–113.

Schliesser, Eric (2011). "Without God: Gravity as a Relational Quality of Matter in Newton's Treatise." In Peter Anstey and Dana Jalobeanu, eds., *Vanishing Matter and the Laws of Motion: Descartes and Beyond*. New York: Routledge, pp. 80–100. To be reprinted in Eric Schliesser (forthcoming), *Newton's Metaphysics*, New York and Oxford: Oxford University Press.

Schliesser, Eric (2013). "The Methodological Dimension of the Newtonian Revolution." *Metascience* 22 (2): 329–333.

Schliesser, Eric and Christopher Smeenk, eds. (forthcoming). *Oxford Handbook of Isaac Newton*. New York and Oxford: Oxford University Press. Selections first published online 2017. DOI: 10.1093/oxfordhb/9780199930418.001.0001

Shapiro, Alan (1993). *Fits, Passions, and Paroxysms: Physics, Method, and Chemistry and Newton's Theories of Colored Bodies and Fits of Easy Reflection*. New York: Cambridge University Press.

Shapiro, Alan (2004). "Newton's 'Experimental Philosophy'." *Early Science and Medicine* 9 (3): 185–217.

Smith, George E. (2005). "Was Wrong Newton Bad Newton?" In Jed Z. Buchwald and Allan Franklin, eds., *Wrong for the Right Reasons*. Dordrecht: Springer, pp. 127–160.

Smith, George E. (2016). "The Methodology of the *Principia*." In Rob Iliffe and George E. Smith, eds., *The Cambridge Companion to Newton*. Second Edition. Cambridge, UK: Cambridge University Press, pp. 187–228.

Snelders, H. A. M. (1989). "Christiaan Huygens and Newton's Theory of Gravitation." *Notes and Records of the Royal Society of London* 43 (2): 209–222.

Wilson, Catherine (2019). "What (Else) Was Behind the Newtonian Rejection of 'Hypotheses'?" In Alberto Vanza and Peter R. Anstey, eds., *Experiment, Speculation and Religion in Early Modern Philosophy*. New York and Oxon: Routledge, pp. 158–183.

Wilson, Curtis (1974). "Newton and Some Philosophers on Kepler's 'Laws'." *Journal of the History of Ideas* 35 (2): 231–258.

Index

Note: Page number followed by "n" refer to footnotes.